GARLAND STUDIES ON INDUSTRIAL PRODUCTIVITY

edited by

STUART BRUCHEY
UNIVERSITY OF MAINE

A GARLAND SERIES

STATISTICAL QUALITY CONTROL IN HIGH RELIABILITY SYSTEMS

LANCE A. MATHESON

GARLAND PUBLISHING, INC.
NEW YORK & LONDON / 1994

Library of Congress Cataloging-in-Publication Data

Matheson, Lance A., 1952–
Statistical quality control in high reliability systems / Lance A. Matheson.
p. cm. — (Garland studies on industrial productivity)
ISBN 0–8153–1628–3 (alk. paper)
1. Quality control—Statistical methods. 2. Quality control—Cost effectiveness—Mathematical models. 3. Flow charts.
I. Title. II. Series.
TS156.M385 1994
658.5'62'015195—dc20 93–34976
CIP

Printed on acid-free, 250-year-life paper
Manufactured in the United States of America

In Memory of Dr. A. R. Matheson

Contents

Tables

Tables

Figures

Preface

This book addresses two key issues in modern manufacturing, the selection of the best statistical quality control charts to use, and the use of buffer inventories after a process to further reduce the cost of quality. The book develops cost minimization algorithms which are applied to the Shewhart c, Cumulative Sum and Geometric Moving Average control charts. The cost performance of these charts is studied as the overall quality of a manufacturing system increases. The c chart is the less expensive chart at relatively low quality levels, and the Cumulative Sum chart surpasses both the c and Geometric Moving Average chart at medium and high quality levels.

Another issue in manufacturing is the disposition of components after they are produced. This is the first book to introduce the use of an inventory buffer, which reduces quality costs, after a production process. The author shows that if there is a cost penalty for assembling defective components into a finished product then an inventory buffer between the production and assembly stations reduces the cost of quality. The book derives the distribution of the quantity produced after a shift in quality and optimality conditions for the buffer size. The original cost model is extended to determine the optimal control chart variables and buffer size for such a production process.

Preface

Acknowledgments

I wish to thank Ted Klastorin for his guidance and support during this research. I would also like to acknowledge the help of Doug Montgomery in the initial research definition and Kamran Moinzadeh and Andrew Siegel in the analysis of this work. Special thanks to Jay Gardiner, Stephen Vardeman, and Amy Boyd for allowing me to use their computer programs.

I am grateful to the Graduate School of Business of the University of Washington for the award of the Edna Benson Fellowship.

Finally, I would like to thank my wife, Jan, and my sons, Bryan and Kevin for all their love and support over the years.

Statistical Quality Control in High Reliability Systems

I. Introduction

1.1 NEW CHALLENGES IN MANUFACTURING

Our world is experiencing a revolution in the way we produce products. The quality of the products has been greatly improved. New production methodologies, such as just-in-time, flexible manufacturing, group technology, and CAD/CAM, allow us to manufacture products at lower cost and at higher quality than ever before. Quality has become a new competitive weapon among companies, encouraging manufacturers to emphasize the quality/reliability of their products in their advertising. As product quality increases, so too do the expectations of the consumer. Customers now expect high quality products which will give them cost-effective service. Management must achieve and maintain these high levels of quality, which requires the selection of appropriate production processes and quality control systems. High reliability production processes and sophisticated test equipment tend to be more expensive than simpler methods, so manufacturers often base the final selection of these production systems on some form of economic analysis.

These multiple requirements, increasing quality, selecting the appropriate technology to maintain it and performing economic analyses, can make use of a quality control method called statistical quality control. Statistical quality control provides tools for improving and maintaining product quality. Also, economic modeling, when applied to statistical quality control, can determine the costs of a quality control system. This data can help us select and justify manufacturing and quality control technology. Before going into the details of the economic modeling, this chapter introduces the concepts of statistical quality control. We also present a motivation for the research and the scope of the book.

1.2 STATISTICAL QUALITY CONTROL

All production systems experience some natural variability in their inputs and transformation processes. No matter how careful the operator is or how well manufacturers maintain the machines, there is always something outside the control of the system (for example, the voltage in a power line fluctuates or a chemical contains some impurities). All of these small, chance causes of variability are "background noise," which causes variability of similar magnitude in the process output. A process which is exhibiting only this natural variability in its output is operating *in statistical control.*

Processes can experience more variability than that attributable to just background noise. When a process undergoes a major change in the quality of its output there is usually a specific, identifiable reason for the change. These major changes usually arise from three sources: improperly adjusted machines, operator errors, and/or defective raw materials (Montgomery, 1985). A process operating under the influence of one of these "assignable causes" is operating *out of control.* These three sources are broad categories and provide flexibility in the classification of causes. We could term the catastrophic failure of a machine an improperly adjusted machine. A major blunder by an operator may be very rare, but it would be an operator error. If a raw material is any consumable item used in the manufacturing process, then a cutting tool which breaks is a defective raw material.

A defect is one commonly controlled quality characteristic. A defect is a point in the product at which a quality specification for the product is not met. Defects might be scratches, dents, or missing components which occur at random throughout the item. Some industries have made great progress in reducing the number of defects in their products to very low levels. Some electronic firms produce so few defects that their occurrence is measured in terms of defects per million units produced (PPM). This has been done, in some cases, by using statistical quality control to identify and reduce the causes of process variability.

Once manufacturers remove the major assignable causes, only small shifts should occur in the process.

Removing those causes of process variability requires management participation. Authors like Deming (1986) and Crosby (1979) have documented the importance of management involvement in improving quality. Management defines, at least implicitly, the acceptable level of variability in a process. As manufacturers identify the causes of variability, they can remove them or allow them to continue to effect the process. To remove the sources of variability requires an investment by management. For example, to reduce blemishes in painted parts might require overhauling the painting system. To maintain a certain tolerance of a machined part might require more expensive cutting tools or a better grade of steel. Reduction of variability might require better operator training and instructions.

In this study, we assume that the number of defects per unit is the quality characteristic of interest and that a Poisson distribution describes the number of defects in a sample. The Poisson approximation to the binomial distribution justifies this. This assumption requires that the number of potential sites for defects in a sample be very large and that the probability of a defect occurring at any site be very small. We can subdivide the *"inspection unit"*, the actual thing which is being inspected, into smaller and smaller sites until this approximation holds. The approximation also requires that the inspection unit be the same for each sample to ensure that the number of sites in each sample is constant. An inspection unit and the related defect could be the number of cracks in 100 feet of weld, the number of blemishes per square feet of painted surface, the number of solder splashes per ten circuit boards, and so on.

We have assumed that manufacturers are monitoring only one quality characteristic. Manufacturers can control multiple variables simultaneously however, such systems are beyond the scope of this book. We also assume that the process has only one out-of-control state. The process shifts to this out-of-control level and stays there until returned to the in-control state. The process cannot repair itself nor return to the in-control state without human intervention. Multiple-state systems, where there are several out-of-control levels of quality, also exist. The modeling of such systems is an extension of the single-state model and requires the estimation of more parameters. Duncan(1956) and Knappenberger and Grandage (1969) have studied multiple-state systems. The single-state model has been more thoroughly studied and is the subject of this research.

When in control, the process produces product at a mean value of μ_0 for the monitored quality characteristic. When a shift occurs, the process mean changes to a new value of

$$\mu_1 = \mu_o \pm \Delta\, \sigma,$$

where Δ is the magnitude of the shift in standard deviations and σ is the standard deviation of the process. Management could choose not to respond to the shift in process quality if it felt that the cost of locating and removing the cause of the shift was more than the cost of repairing the non-conforming items. By reacting to the shift, we are defining a discrete control system; the process is operating at an acceptable or an unacceptable level. By definition, a quality level between μ_o and μ_1 is also acceptable because no response is made until the process reaches the μ_1 level.

Statistical quality control rests on the premise that we can monitor the quality of a process by taking samples from the process at pre-specified time intervals. Each sample is a test of the hypotheses

$$H_0 : \mu = \mu_o$$
$$H_1 : \mu \neq \mu_o\,,$$

where μ_0 is the average value of the quality characteristic which is being controlled and μ is the hypothetical average value. If we do not accept the null hypothesis then we assume that a shift has occurred. The detection of a shift in quality causes us to examine and readjust the process. This periodic sampling and hypothesis testing requires the determination of a sample size, sampling interval and confidence level for the test.

A record of these hypothesis tests would allow us to monitor process performance over time. A *control chart* combines this recording function with a time series display of each hypothesis test. The control chart uses control limits to indicate the confidence interval for the hypothesis test. The centerline, *CL,* of the control chart represents the average, in control, value, μ_0, of the quality characteristic for the process. The upper control limit , *UCL,* of the control chart is the largest value for the sample statistic at which we would accept the null hypothesis. The lower control limit , *LCL,* of the control chart is the lowest value for the sample statistic at which we would accept the null hypothesis. The values of the upper and lower control limits are the corresponding limits of the confidence interval for the process mean. As long as the sample statistics fall between these control limits, we say the process is in control. When a sample statistic exceeds the control limits, we assume that the process has gone out of control. When this occurs, we further assume that an assignable cause exists and corrective action is necessary to find and remove the cause.

There is a chance that we will make the wrong decision when we test a hypothesis. If we reject the null hypothesis when it is true, then a Type I error has occurred. If we do not reject the null hypothesis when it is false, then a Type II error has occurred. The probabilities of these two types of errors are

$$\alpha = \text{p(Type I error)} = \text{p(reject } H_o \mid H_o \text{ is true)}$$
$$\beta = \text{p(Type II error)} = \text{p(fail to reject } H_o \mid H_o \text{ is false).}$$

If a Type I error occurs, we conclude that the process has shifted when it has not (a false alarm). On the other hand, if we commit a Type II error, then we fail to detect the shift, and the process continues to operate out of control. The sample size, magnitude of the shift and confidence level for the test determine the values of α and β. Ideally, we want to design a sampling scheme that does not give many false alarms but that quickly detects a real shift in quality. The expected number of samples taken before a false alarm occurs is the average run length in control, *ARL1*. The expected number of samples required to detect an out of control condition is the average run length out of control, *ARL2*. The mathematical derivation of ARLs appears in Chapter III.

1.3 MOTIVATION FOR WORK

Heuristic approaches have traditionally determined the values of the decision variables for a statistical quality control chart. Given a value of α, we can compute β given a sample size and the magnitude of the shift, Δ. Once we know values of α and β, we can calculate the ARLs and choose a sampling interval. Often, the quality engineer chooses the sampling interval based on experience. This set of decision variable values is called the *design of the chart.* If we are trying to choose the "best" design from among several possible designs, what criteria should we use? The ARLs provide a measure of the control chart's ability to detect a shift. The sampling interval indicates how often to sample, while the sample size equals the number of items to take. There are few rules on how to compare these different measures to arrive at a "good" design. A lower value of ARL1 for one design might require a larger sample than some other design which makes it unclear which design is really better.

Managers and engineers who design statistical quality control charts can consider the financial implications of their decisions. Taking samples from the process costs time and money, so the sampling frequency and the size of the samples have a direct impact on the profitability of a process. This is even more important when we consider expensive test procedures, like destructive testing. Searching for the cause of defective items can be costly, especially if the operator must shut down the process before a search or repair begins. This idle time means lost production and revenue. The values of α and β determine the frequency of out-of-control signals which affects the amount of time and money spent searching for assignable causes. The consideration of cost is implicit in the determination of the values for these decision variables.

Management and quality control engineers make implicit financial decisions about the cost of quality when they design production systems and control schemes. We assume, therefore, that the cost of quality control is the appropriate measure of performance and that it is superior to the performance measures used in the heuristic procedures. Minimizing the cost of quality control should be the explicit objective of the design process. Therefore, to minimize the cost of quality, we must determine the values for the decisions variables, which yield the lowest cost design.

Researchers have developed a number of economic models to quantify the costs of quality control. They determine the lowest cost design for a particular set of process parameters. We denominate the objective function value of these models in dollars per unit time which allows a direct comparison of different control charts. This eliminates the need for the comparison of other performance measures, which we might not as easily interpret.

We can use different statistics to evaluate the performance of a process but researchers often use the sample mean, the accumulated deviations from a target value and the weighted average of the observations. Economic models allow any of these statistics to be the basis of the model. There have not been any studies of the effect of the choice of statistic on the cost of quality control. The introduction of higher quality manufacturing systems raises new questions. What happens to the values for α and β when the magnitude of the shift gets very small? What is a good sampling interval when the average time between process shifts is on the order of two to three weeks?

1.4 THE ALLOCATION OF INSPECTION EFFORT PROBLEM

This work considers the design of a quality-control chart for a single machine or operation. If we extend this work to the design of control charts for an entire production system, the analysis might benefit from considering any dependencies between machines. The study of the allocation of inspection effort to a production line provides a starting point for this extension. The allocation of inspection problem determines how many inspection sites to locate along a production line and what proportion of items to inspect at each site. We review the significant findings in this field and show the relationship to statistical quality control.

Lindsay and Bishop (1964) study a serial flow process where the process produces too many defective items to meet the desired quality level. Therefore, manufacturers must screen the process output to remove enough defective items to attain the desired quality level. They model a single-line, multi-stage process where each stage has a constant probability of producing a defective item. They know the costs of inspecting and scrapping an item at each stage. The objective is to find the proportion of items to inspect at each stage which minimizes the total cost of inspection and scrap subject to a requirement on product quality. They determine that the optimal inspection policy is to inspect all or none of the production at any stage. They further show that inspection is performed at only a subset of the possible inspection points.

White (1968) studies a serial batch process and determines the optimal number of items to inspect in each batch. He found that either all or none of a batch is inspected at a subset of the potential inspection sites. Britney (1972) extends the Lindsay and Bishop work to non-serial production systems with the same zero or 100 percent inspection result.

Yum and McDowell (1981) study a non-serial production system in which the product of any operation is defective if it has at least one defective component or if it has acquired a defect as a result of processing. They do not place a constraint on product quality but do add a cost for external failures. Their objective is to minimize the sum of inspection, repair and external failure costs subject to the requirement that the amount of inspection at each site be between 0 and 100 percent. They find that a none or all inspection policy is optimal.

There are two major differences between statistical quality control and the allocation of inspection: the number of production stations and the probability of producing a defective item. Statistical quality control deals with monitoring the performance of a single operation. It assumes that the probability of producing a defect , when in-control, is acceptable. Therefore, if the process remains in control, no inspection is necessary.

Eventually the defect rate of the process shifts to an out-of-control level which is not acceptable. We want to detect this shift in process quality and restore the process to the in-control state. We need some form of inspection to detect this shift. Because SQC uses periodic samples, it implies that periodic sampling is less expensive than 100 percent inspection. If the process is in control most of the time, then 100 percent inspection would mainly inspect acceptable items. If operators do not inspect the process, we would fail to detect the shift and allow production to proceed at the unacceptable level. The design of the control chart attempts to minimize the sum of inspection, out-of-control production and process repair costs for one station.

The allocation of inspection literature assumes that the probability of producing a defective item at each stage is fixed, so detecting a shift in quality is not an issue. The 0-100 percent inspection solution implies that, at each station, if the cost of passing a defective item is less than the cost of inspecting the item, do not inspect it. If we apply this result to a one-station line, the relationship to statistical quality control becomes apparent. Manufacturers use inspection in statistical quality control because the process might go out of control, and the cost of defects would become too high. Manufacturers use inspection in the allocation problem at those sites where the cost of defects is too high. Thus, statistical quality control can be thought of as a one-station allocation of inspection effort problem with two probabilities of producing a defective item.

Combining the assumptions of these problems and applying them to the allocation and economic design of statistical quality control charts for a serial production process would be a useful extension of quality control.

1.5 SCOPE OF WORK

The purpose of this book is to study the ability of an economic model for statistical quality control to determine the lowest cost control chart for a particular process. We apply the economic model to systems where the number of defects per unit is the quality characteristic of interest. We also investigate the relative cost performance of these control charts as the process quality improves. We present new algorithms to solve the optimization problems involved in this process and extend the concept of economic design of a control chart to the determination of the optimal buffer inventory for a process.

Chapter II continues the explanation of statistical quality control. We review the literature on the economic design of quality control charts and the calculation of ARLs and explain the relationship between statistical quality control and inspection. We develop the economic model for this research from the literature.

Chapter III presents the mathematical properties of the economic model and the algorithms required to optimize it. We introduce the models for computing ARLs and use a numerical example to demonstrate the selection process. We compare the economic performance of three control charts at different levels of quality.

Another issue in manufacturing is the disposition of components after production. If manufacturers need to repair these parts where should they store the parts while they await repair? Manufacturers often immediately assemble items into a product. This means that they assemble defective units before identifying and repairing them. A buffer inventory between the machine and assembly stages reduces repair costs if it is less expensive to repair the component parts before assembly. The repair costs savings offset increases in inventory holding costs. This is the first work to discuss the use of buffer inventory in quality control. Chapter IV extends the economic model developed in Chapter II to the design of production systems which use buffer inventory. We develop the distribution of the number of units produced after a shift in quality and derive optimality conditions for the buffer inventory. A numerical example demonstrates the procedure.

II. The Economic Model For Control Charts

2.1 OVERVIEW

This chapter continues the explanation of statistical quality control and describes the economic design of control charts. The next section describes the three types of control charts used in this study, Shewhart, CUSUM and GMA. Section 2.3 reviews the literature on the economic design of quality control charts and the calculation of average run lengths. We develop the cost model for this research from the literature in section 2.4.

2.2 STATISTICAL QUALITY CONTROL CHARTS REVISITED

Chapter I introduced quality control charts which provide a visual display of the hypothesis tests of process control over time. The centerline of the chart is set at the average, in control, value for the process. The upper and lower control limits are the limits of the decision interval for the hypothesis test. The hypothesis test and the design of the quality control chart rely on a set of variables: the sample size n, sampling interval h, and desired sensitivity to changes in the quality level of the process L *or* L_{cu}.

There are three basic types of control charts: Shewhart, Cumulative Sum (CUSUM) and Geometric Moving Average (GMA). These charts appear in Figures 2.1 to 2.3, while their respective variables are in Tables 2.1 to 2.3, respectively.

Table 2.1 Shewhart Control Chart Notation

Notation	Meaning
n	sample size
h	sampling interval
L	sensitivity factor
UCL	upper control limit
LCL	lower control limit
μ_0	in-control process mean
σ_n	standard error of process or standard deviation of mean
CL	centerline = μ_0

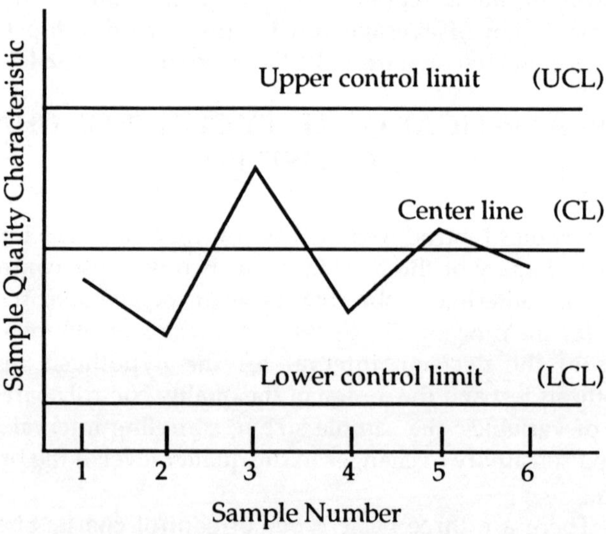

Figure 2.1 Shewhart Quality Control Chart

Table 2.2 CUSUM Control Chart Notation

Notation	Meaning
n	sample size
h	sampling interval
k	CUSUM target value
L_{cu}	CUSUM decision interval
d	lead distance
Θ	V-mask angle
μ_0	in-control process mean

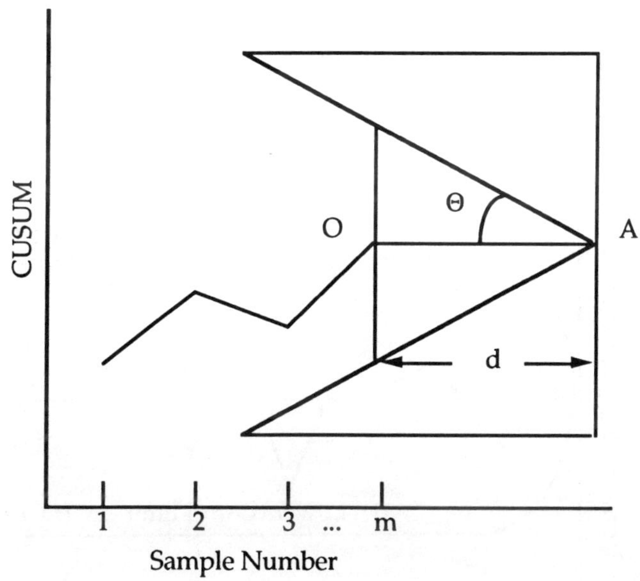

Figure 2.2 CUSUM Quality Control Chart
With V-mask

Table 2.3 GMA Control Chart Notation

Notation	Meaning
n	sample size
h	sampling interval
L	sensitivity factor
r	weighting factor
UCL	upper control limit
LCL	lower control limit
μ_0	in-control process mean
σ_n	standard error of process or standard deviation of mean
σ_{gma}	standard deviation of moving average
CL	centerline = μ_0

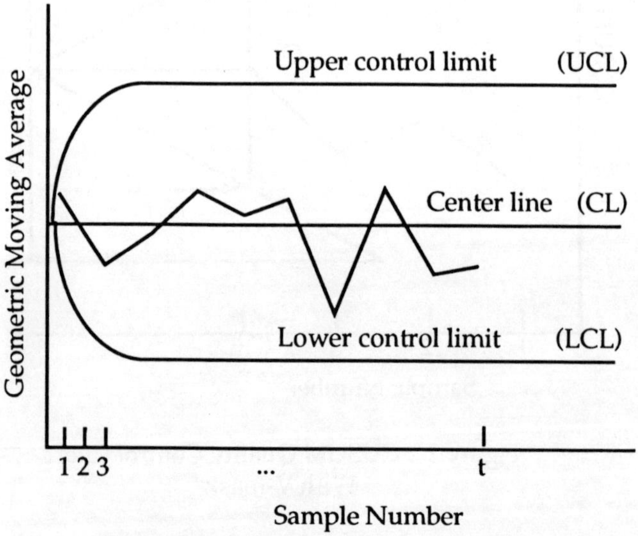

Figure 2.3 GMA Quality Control Chart

2.2.1 The Shewhart Quality Control Charts

The first control chart, a Shewhart chart, bears the name of Walter Shewhart of Bell Telephone Laboratories (Shewhart, 1926 a, b). He developed a family of control charts to control a number of different statistics to describe the quality of a process. The " $\bar{x}$ chart" tracks the average of sample measurements like weight or height. The "R chart" controls the range of measurements within a sample. The "p chart" monitors the proportion of defective items in a sample. The "c chart" records the count of defects in a sample.

The control limits of a Shewhart chart are

$$UCL = \mu_0 + L\,\sigma_n$$
$$\text{Centerline} = CL = \mu_0$$
$$LCL = \mu_0 - L\,\sigma_n\,,$$

where μ_0 is the in-control process average, σ_n is the standard error of the process and L is the desired sensitivity level. In the heuristic determination of values for the control chart, L has usually been set at a value of three. This means that we are setting the limits on our confidence interval, for the average, at $\pm 3\sigma_n$. Assuming a normally distributed random variable, this corresponds to a 0.997 probability that the confidence interval will contain the true mean. This is the same as saying that there is a 99.7 percent chance that the sample point will plot within the control limits when the process is in control if the statistic follows the normal distribution.

The variable L represents the desired sensitivity of the control chart to shifts in quality, and the parameter Δ is the magnitude of the shift we want to detect. We choose a value for L which will detect a change of magnitude Δ in an acceptable period of time. If we were to let $L = \Delta$ then, after a shift occurs, the sample means would exceed the control limits only 50 percent of the time. If there is only a 50 percent chance that a sample mean will exceed the control limits, it will take longer to detect the shift then if L is greater than Δ.

It is possible to compute a negative value for the lower control limit if the standard error is relatively large compared to the mean or if the mean is very small. If this happens we can set the lower control limit equal to zero, and a sample statistic value of 0 would be in-control. We use this modification frequently for the c chart which monitors the number of defects in a sample. Since we usually monitor an increase in the number of defects as detected by the upper control limit, this modification of the lower limit is insignificant.

We should note that there are systems in which we are also interested in detecting decreases in the reported number of non-conforming units. When a sample point plots below the lower control limit of a c or p chart, we assume that there has been an increase in the quality of the process. We want to know if this is a real improvement or the result of a breakdown in the system. For example, if a test machine goes out of adjustment, we might not detect the non-conforming units and would find fewer defects in a sample. We want to know if the test equipment is defective or if the process has actually improved.

2.2.2 The Cumulative Sum Chart

Page (1954) developed the Cumulative Sum, CUSUM, chart. A CUSUM chart plots the cumulative sum of the deviations of the sample statistics for any quality characteristic from the target value *k*. The mth cumulative sum is

$$S_m = \sum_{i=1}^{m} (x_i - k),$$

where x_i is the statistic of the ith sample. The target value for the process, *k*, reduces the magnitude of the cumulative sum and makes changes in the size and direction of the sum easier to recognize. Common heuristic choices have been the in-control process mean, μ_0, or the reference value for a sequential probability ratio test:

$$k = \frac{\mu_1 - \mu_0}{\ln(\mu_1) - \ln(\mu_0)} \text{ ,}$$

where μ_1 is the out-of-control process mean (Lucas 1985). Table 2.2 defines the *decision limit* for the CUSUM chart to be L_{cu}; that is, if a plotted point rises a distance L_{cu} or more above the lowest previous point, we assume that there has been an upward shift in the process. If a point falls a distance $-L_{cu}$ or more below the highest previous point we assume that there has been a downward shift in the process. The value of L_{cu} is a function of k and the desired average run length for the control chart. The choice of k determines the magnitude of the cumulative sum and therefore the value of L_{cu} needed to indicate an out-of-control condition. Thus the sample size n, sampling interval h, target value k, and decision limit L_{cu} are the variables required for operating a CUSUM chart.

A CUSUM chart is an application of Wald's sequential probability ratio test (SPRT) to the sequence $S_m, S_{m-1}, S_{m-2}, \ldots,$ starting with the mth observation and moving backwards (Johnson and Leone 1962, Khan, 1984). Researchers use the SPRT to test the null hypothesis, H_0, that the process remains in control

$$H_0 : \mu = \mu_0$$
$$H_1 : \mu \neq \mu_0.$$

The likelihood for a distribution, given m samples, is

$$L_{0,m} = \prod_{i=1}^{m} f(x_i \mid \mu_0),$$

where $f(x_i \mid \mu_0)$ is the distribution function of x_i under the null hypothesis. Let α be the probability of a type I error and β be the probability of a type II error. For a given sum of m sample observations, we can compute the likelihood ratio of $\frac{L_{1m}}{L_{0m}}$ and subject it to the following test:

i) if $L_{1m}/L_{0m} \leq \beta/(1-\alpha)$, we accept H_0
ii) if $L_{1m}/L_{0m} \geq (1-\beta)/\alpha$, we reject H_0.
iii) if $\beta/(1-\alpha) < L_{1m}/L_{0m} < (1-\beta)/\alpha$, we take a further observation, compute the sum S_{m+1} and calculate the new likelihood ratio $\frac{L_1(m+1)}{L_0(m+1)}$.

We repeat this process until we reach a decision.

Figure 2.4 shows the test regions of a CUSUM chart, where we are plotting the mth cumulative sum versus the sample number, m. The angle Θ, the *lead distance* d, and the decision interval L_{cu} define the acceptance and rejection regions, where

$$d = \left(\frac{2}{\Delta^2}\right) \ln \left(\frac{1-\beta}{\alpha}\right)$$

and

$$\Theta = \tan^{-1}\left(\frac{\mu_1-\mu_0}{2k}\right)$$

where

$$\Delta = \frac{\mu_1-\mu_0}{\sigma_n}$$

is the magnitude of the shift, in standard errors, that the chart is to detect. Here α is the probability of falsely concluding that a shift has occurred, β is the probability that the shift will go undetected, μ_1-μ_0 is the change in means which we want to detect and σ_n is the standard deviation of the process mean. If the cumulative sum of the ith sample falls between the sloping lines then we take another sample. If the cumulative sum falls within the "accept H_0" region, then accept the hypothesis that the process is in control. If the cumulative sum falls in the "accept

H_1" region, then accept the hypothesis that the process is no longer in control.

A piece of clear material can be cut in the shape of the test regions and laid over the plot of the cumulative sums. Barnard (1959) proposed this visual aid, called a V mask, to simplify the decision process. Figure 2.2 shows a V mask, where the point O is placed on the most recent cumulative sum and the line OA is parallel to the horizontal axis. If one of the previously plotted points falls outside of the arms of the V mask, then we reject the null hypothesis that the process is in control.

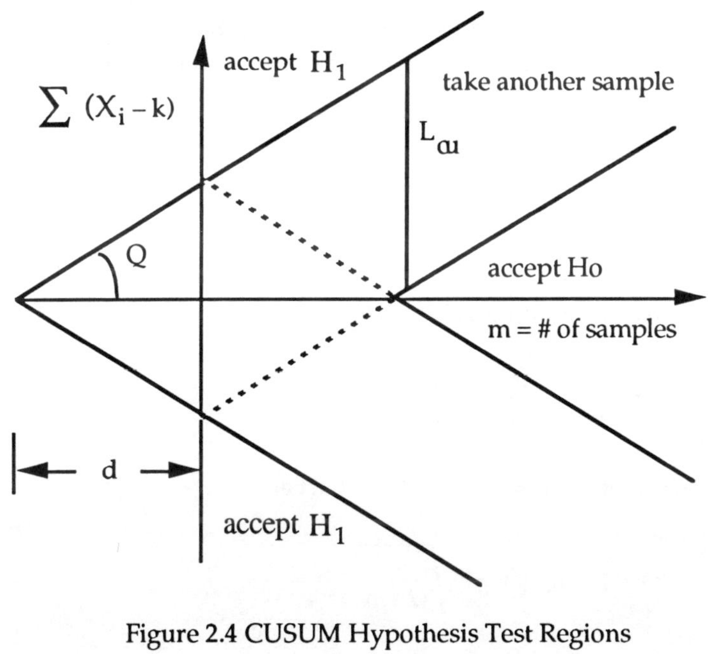

Figure 2.4 CUSUM Hypothesis Test Regions

2.2.3 The Geometric Moving Average Chart

Like the cumulative sum, the geometric moving average (GMA) incorporates prior data in the plotted statistic. Roberts (1959) developed the GMA chart based on the geometric moving average of the sample averages. The variable plotted on the chart is

$$Z_{t+1} = rX_t + (1-r)\,Z_t$$

where $0 \leq r \leq 1$ is a weighting constant, X_t is the sample statistic at time t, Z_t is the predicted moving average at time t and the starting value of Z_0 is the long run process mean μ_0. Hunter (1986) states that experience with econometric data suggests values of $r = 0.2 \pm 0.1$, but he gives no justification for these values. Plotting the values of Z_{t+1} on the control chart at time t, gives a forecast of where the process will be at the next sample point. If we assume that the X_t are independent random variables with variance $\frac{\sigma^2}{n}$, then the variance of Z_t, σ^2_{gma} is

$$\sigma^2_{gma} = \frac{\sigma^2}{n}\left(\frac{r}{2-r}\right)[1-(1-r)^{2t}]$$

which has a limiting value, as $t \to \infty$, of

$$\sigma^2_{gma} = \frac{\sigma^2}{n}\left(\frac{r}{2-r}\right).$$

The control limits for the GMA chart are

$$UCL = \mu_0 + L\sigma\sqrt{\frac{r}{(2-r)n}}$$

$$LCL = \mu_0 - L\sigma\sqrt{\frac{r}{(2-r)n}},$$

where L is the sensitivity factor of the control chart. Figure 2.3 shows the curved control limits which become parallel as σ^2_{gma} approaches its limiting value.

2.2.4 Relationships Between Control Charts

While these control charts appear unrelated, we can view them as a continuum of models. The charts use the same sample statistics as their inputs but assign different weights to them over time. The Shewhart chart puts all of the weight on the most recent observation, ignoring prior data. The weight given the last plotted point, X_t, is $w_t=1$, and the weight given to prior observations is $w_{t-k}=0$, for $k\geq1$. The CUSUM chart accumulates m sample statistics in the mth sum, with each term having equal weight, $w_t=1/m$, for all t. The GMA chart gives decreasing weights to each term in the series

$$Z_t= rX_t + (1-r)\,Z_{t-1}\,,$$

where $w_i = r(1-r)^{t-i}$. Figures 2.5 through 2.7 show the weightings for each of these charts. The exponential weighting of the GMA chart explains the relationship between these three charts. The weighting parameter *r* determines the rate of decay of the weights and hence, the amount of information provided by prior observations. As *r* approaches 1, $w_t=r(1-r)^{t-i}$ approaches 1, and the most recent observation is almost fully weighted, the same as the Shewhart chart. As *r* approaches 0, $w_t=r(1-r)^{t-i}$ approaches 0, the weights of the current observation and the earlier observations are almost equal (the same as the CUSUM chart) Thus, the CUSUM, GMA and Shewhart charts are a range of charts which use all, part and none of the prior information, respectively, about a process. While the charts treat the data differently, they all work to detect shifts in process quality.

However, the amount of retained information affects the control charts responsiveness to shifts in process quality. The CUSUM chart is more effective than the Shewhart chart in detecting small shifts in process quality. However, the CUSUM does not detect large shifts as quickly as the Shewhart chart (Montgomery, 1985). It should be possible to determine the optimal weighting for the GMA chart given the magnitude of the process shift.

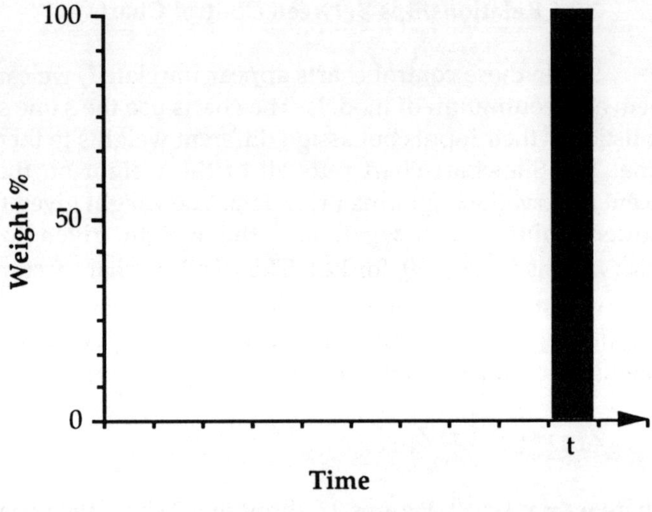

Figure 2.5 Data Weighting for the Shewhart Chart

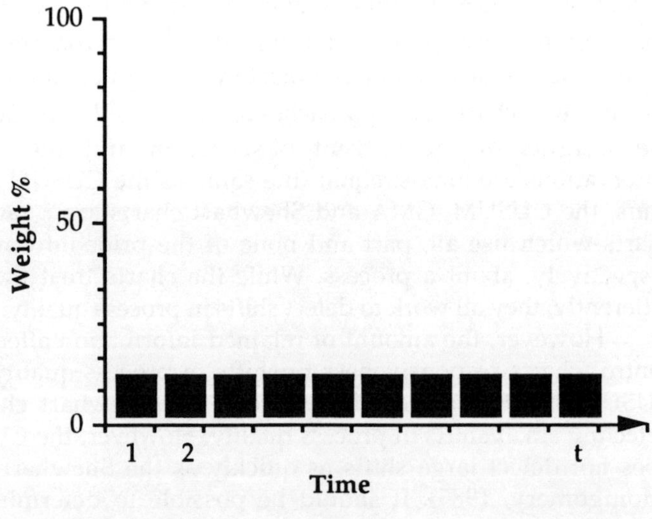

Figure 2.6 Data Weighting for the CUSUM Chart

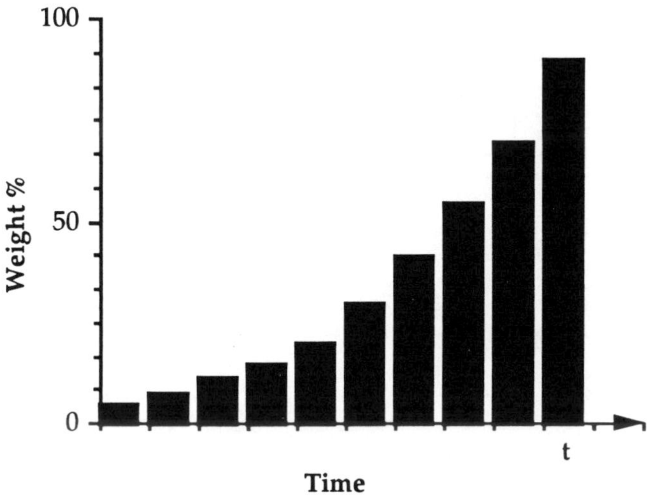

Figure 2.7 Data Weighting for the GMA Chart

2.3 THE ECONOMIC DESIGN OF QUALITY CONTROL CHARTS

Define a quality control cycle as the time between the start of successive in-control periods. The cycle begins with the process operating in control. The process then shifts to a higher defect rate; the operator detects the shift and returns the process to the in-control state. This cycle defines a renewal-reward process. Renewal theory implies that the expected cost per cycle divided by the expected cycle time defines the expected cost per time period of this process (Ross, 1970). Thus, given an expected value of the cycle time E(T) and an expected cost per cycle E(C), the expected hourly cost, E(A), is merely $\frac{E(C)}{E(T)}$.

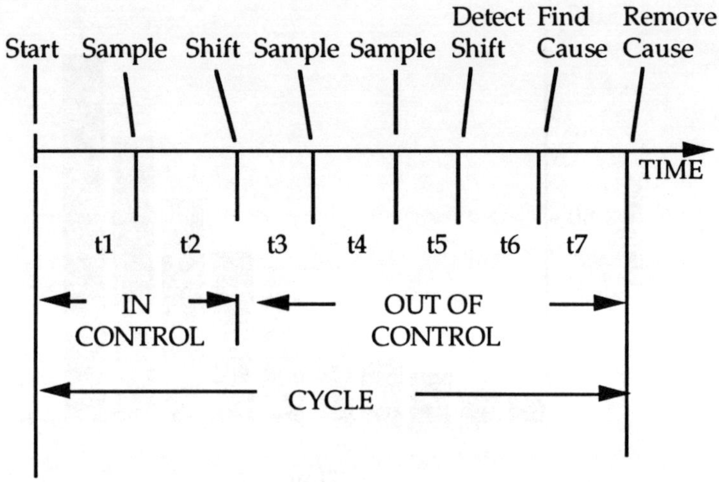

Figure 2.8 Diagram of the Quality Control Cycle

We assume that the process has only one out of-control-state and that it cannot repair itself or return to the in-control state without human intervention. There are models for multiple-state systems which are extensions of the single-state model and require the estimation of more parameters. A multiple-state system could experience more than one shift in quality between the initial shift and the detection of the out-of-control condition. We assume that these systems cannot repair themselves and that each successive shift decreases the quality of the process. We can expand Figure 2.8 to depict a multiple-state system by inserting more shifts in quality during the times t3 through t5. The literature on multiple-state systems assumes that the time until the first shift follows an exponential distribution with mean $\frac{1}{\lambda}$ which is the same assumption as in the single-state systems. We assume that the shifts between levels of process quality define a Markov process with known transition probabilities between states. Duncan (1956) and Knappenberger and Grandage (1969) state that a carefully designed single-state approximation to a multiple-state system gives good results.

Researchers have more thoroughly studied the single state model and we made it the subject of this research.

The initial work on the economic design of control charts dealt with the $\bar{x}$ chart. Girshick and Rubin (1952) developed a minimum cost control procedure for the case in which the time to breakdown had a geometric distribution. They were the first authors to use the expected cost per unit time criterion. Their optimal procedure requires the solution of complicated integral equations and hence has found little acceptance in practice.

Several early researchers developed models for Shewhart charts which were either incomplete or did not apply formal optimization procedures to the model. Montgomery (1980) refers to these studies as semi-economic designs. Weiler (1952) sought to minimize the total amount of inspection required to detect a shift and did not formally consider costs. Taylor (1968) did the early investigation of the economic design of CUSUM charts for normally distributed variables. He expressed the cost per unit time as a function of the sample size n , sampling interval h , and the V-mask design variables d and θ, but he assumed that n and h were pre-specified in order to optimize the model. Cowden (1957), Barish and Hauser (1963), and Aroian and Levine (1950) have proposed other semi-economic models.

Duncan (1956) developed the first model for the total economic design of a Shewhart control chart, and he developed optimization procedures to determine the optimal chart variables. Like Girshick and Rubin, Duncan sought to maximize the expected net income per unit time as a function of the variables (sample size, sampling frequency and sensitivity factor) of a Shewhart chart. Duncan's work has been the stimulus and basis for much of the work on the economic design of control charts which followed. Goel and Wu (1973) developed an economic design model for the one-sided CUSUM chart with normally distributed quality characteristics. Chiu and Wetherhill (1974) developed a semi-economic procedure for the design of an $\bar{x}$ chart.

Duncan assumed that the process continued to operate after the chart indicated a shift in quality. Several authors have modified the model to allow for process shutdown. Chiu (1974) developed an economic model of the CUSUM chart where production stopped to allow searching for the cause of the shift. Panagos, Heikes and Montgomery (1985) revised Duncan's model to study the impact of stopping the process to search for the cause and to repair the process. They show the importance of choosing the right cost model for the process. Lorenzen and Vance (1986) add two binary variables to the Duncan model, which allows it to be applied to continuous or discontinuous production systems.

2.3.1 Average Run Lengths

A number of authors have studied the calculation of average run lengths for CUSUM and GMA charts. Page developed an expression for the CUSUM average run length in his 1954 article. The calculation of these average run lengths requires numerical integration, so before the wide availability of computers, several authors published tables and graphs of average run lengths for CUSUM charts. An engineer could use these published data to design a control chart for his system or to get an estimate of the performance of his design. Goldsmith and Whitfield (1961) used simulation to compute average run lengths of CUSUM charts for normal means. Kemp (1961) shows how to design the CUSUM V mask to obtain a desired average run length for a limited set of parameters. Goel and Wu (1971) produce contour nomograms which allow estimates of average run lengths for CUSUM charts. Brook and Evans (1972) developed a Markov chain approximation for CUSUM average run lengths applicable to continuous or discrete variables. This has proven to be a particularly useful approximation and has been used by Lucas (1985) and Gardiner (1987) to study run lengths for CUSUM charts. Waldmann (1976) and Woodall (1984) developed bounds on the distribution of CUSUM average run lengths. Vardeman and Ray (1986) calculate average run lengths for Poisson and exponential CUSUM charts via numerical integration.

The GMA chart has not received as much attention with respect to average run lengths. Roberts (1959 and 1966) used simulation to compute average run length values for the GMA chart. Robinson and Ho (1978) view the problem of determining average run lengths as a special case of constructing distribution functions for times to first passage of autoregressive processes across ruin barriers. Crowder (1986) published a computer program which evaluates the integral equations which define the average run lengths for GMA charts. All of these GMA articles deal with normal variables. Gardiner (1987) extended the work of Brook and Evans (1972) to the case of GMA charts with Poisson and Neyman-A variables.

2.4 FORMULATION OF COST MODEL

The quality control cycle modeled appears in Figure 2.8. This is the cycle which Duncan and all subsequent authors have modeled. Table 2.4 shows the notation used in the cost model.

2.4.1 Expected Cycle Time Development

Duncan (1956), and subsequent authors, assumed a Poisson process with rate λ generates assignable causes. The time until the process shift occurs is then distributed exponentially with mean $\frac{1}{\lambda}$. This assumption is significant because it allows us to make use of the memoryless property of the exponential distribution, and it makes the mathematics tractable. The empirical studies of Davis (1952) and Epstein (1958) support this assumption. The length of the in-control portion of the cycle in Figure 2.8 is t1 + t2. If the process continues to operate during the search for the cause of the shift, then the expected value of t1+ t2 is $\frac{1}{\lambda}$. If the process stops when the manufacturer conducts a search this postpones the occurrence of the assignable cause by the idle time induced by the false alarms.

Table 2.4 Parameters of Cost Model

Parameters of Cost Model	
Times (hrs)	
Between process shift	$\frac{1}{\lambda}$
Sample/chart one item	E
Search for false alarm	T_0
Search for real alarm	T_1
Repair process	T_2
Costs ($)	
Repair of a defect	D
False alarm	Y
Real alarm	W
Fixed sampling	a
Variable sampling	b
Miscellaneous	
Magnitude of shift (std dev)	Δ
Mean # of defects per unit	
in-control	Λ_0
out-of-control	Λ_1
Production during search	δ_1
Production during repair	δ_2
Production rate (units / hr)	Θ
E(#samples taken in-control)	s
E(occurrence time of shift after sample)	τ

This expected idle time is a function of the expected time needed to search for a false alarm, T_0, and the number of false alarms in the cycle. The number of false alarms is a function of the number of samples taken during the in-control portion of a cycle. Let *s* be the number of samples taken while the process is in control. Then:

$$s = \sum_{i=0}^{\infty} i \Pr\begin{pmatrix}\text{assignable cause occurs between} \\ \text{the ith and (i+1)st samples}\end{pmatrix}$$

$$= \sum_{i=0}^{\infty} i \int_{ih}^{(i+1)h} \lambda e^{-\lambda t} dt$$

$$= \sum_{i=0}^{\infty} i[e^{-\lambda hi} - e^{-\lambda h(i+1)}]$$

$$= (1-e^{-\lambda h}) \sum_{i=0}^{\infty} ie^{-\lambda hi}$$

$$= -(1-e^{-\lambda h}) \frac{d}{d(\lambda h)} \sum_{i=0}^{\infty} e^{-\lambda hi}$$

$$= -(1-e^{-\lambda h}) \frac{d}{d(\lambda h)} \left(\frac{1}{1-e^{-\lambda h}}\right)$$

$$= (1-e^{-\lambda h}) \frac{e^{-\lambda h}}{(1-e^{-\lambda h})^2}$$

$$= \frac{e^{-\lambda h}}{1-e^{-\lambda h}} \quad .$$

If we let

$$\delta_1 = \begin{cases} 1 \text{ if production continues during search} \\ 0 \text{ otherwise,} \end{cases}$$

then the expected time until the process shift occurs is

$$\frac{1}{\lambda} + \frac{(1-\delta_1)\, s\, T_0}{ARL1},$$

which is the expected value of (t1+ t2).

There are t2 hours between the last sample prior to the shift and the occurrence of the shift. Let τ =E(t2) represent the expected occurrence time of the shift when we take samples every h hours and the shift occurs in the interval between the ith and the $(i+1)^{st}$ sample. Then

$$\tau = \frac{\int_{hi}^{h(i+1)} \lambda(x-hi)e^{-\lambda x}\,dx}{\int_{hi}^{h(i+1)} \lambda e^{-\lambda x}\,dx}$$

$$= \frac{\frac{1}{\lambda} e^{-\lambda x}(1+\lambda x) + (hi)\, e^{-\lambda x} \Big|_{hi}^{h(i+1)}}{e^{-\lambda x} \Big|_{hi}^{h(i+1)}}$$

$$= \frac{1-(1+\lambda h)e^{-\lambda h}}{\lambda(1-e^{-\lambda h})}.$$

Then E(t3) = h- τ is the expected time remaining between the shift and the next sample. After the process shift occurs, sampling continues for t4 hours until we take a sample which contains enough defects to give an out-of-control signal. The expected number of samples required to detect a shift after it occurs is ARL2. There are then (ARL2-1) sampling intervals between the occurrence and detection of the shift. The expected number of hours between the shift and the sample which allows detection is h(ARL2-1).

It takes t5 hours to evaluate a sample and plot the results. If sampling and inspection takes E hours per item, then a sample of n units requires t5 = nE hours.

The time to search for an assignable cause is t6 which has an expected value of T_1. It takes t7 hours to repair the process with an expected value of T_2. The expected in-control portion of the cycle is

$$\frac{1}{\lambda} + (1-\delta_1)\frac{sT_0}{ARL1}, \tag{2.1}$$

and the expected out-of-control portion is

$$-\tau + nE + h(ARL2) + T_1 + T_2. \tag{2.2}$$

The expected cycle time is the sum of (2.1) and (2.2),

$$\frac{1}{\lambda} + (1-\delta_1)\frac{sT_0}{ARL1} - \tau + nE + h(ARL2) + T_1 + T_2 \tag{2.3}$$

2.4.2 Expected Cycle Cost Development

We have assumed throughout this work that the process produces defects at a rate of Λ_0 defects per unit from an underlying Poisson process. Let D represent the cost per defect and Θ the production rate in units per hour. Then $C_0 = \Lambda_0 \Theta D$ is the expected hourly cost of defects at the in-control rate. The process shifts to a higher defect rate, $\Lambda_1 = \Lambda_0 + \Delta\sqrt{\Lambda_0}$, where Δ is the magnitude of the shift in standard deviations. The expected hourly cost of defects out-of-control is now $C_1 = \Lambda_1 \Theta D$, with $C_1 > C_0$.

We expect the process to operate in control for $\frac{1}{\lambda}$ hours, so $\frac{C_0}{\lambda}$ is the expected cost of defects produced while in control. If we let

$$\delta_2 = \begin{cases} 1 \text{ if production continues during repair} \\ 0 \text{ otherwise,} \end{cases}$$

and if we use δ_1 as before, then the expected operating time out of control is

$$-\tau + nE + h(ARL2) + \delta_1 T_1 + \delta_2 T_2. \tag{2.4}$$

The expected cost of defects during the out-of-control portion of the cycle is C_1 times (2.4)

$$C_1(-\tau + nE + h(ARL2) + \delta_1 T_1 + \delta_2 T_2). \tag{2.5}$$

The cost per false alarm is Y, which includes the cost of downtime if we stop the process during the search. W is the cost of finding a real cause and repairing the process. Most authors assume that inspection and sampling costs consist of a fixed cost per sample, *a*, and a per unit cost, *b*. Therefore, the cost of sampling and inspecting each sample is a + bn, where *n* is the sample size.

The expected number of samples in a cycle is the ratio of total expected production time to sampling interval or

$$\frac{\frac{1}{\lambda} - \tau + nE + h(ARL2) + \delta_1 T_1 + \delta_2 T_2}{h}.$$

Each sample costs (a + bn) dollars so the expected sampling cost per cycle is

$$(a + bn)\left(\frac{\frac{1}{\lambda} - \tau + nE + h(ARL2) + \delta_1 T_1 + \delta_2 T_2.}{h}\right). \tag{2.6}$$

Since we expect $\frac{S}{ARL1}$ false alarms per cycle, the expected cost of false alarms per cycle is

$$\frac{YS}{ARL1}. \tag{2.7}$$

There is only one real shift in each cycle, so we spend W dollars finding and removing the assignable cause.

Combining these cost terms (2.5 - 2.7) with W gives the expected cost per cycle:

$$\frac{C_0}{\lambda} + C_1[-\tau + h(ARL2) + nE + \delta_1 T_1 + \delta_2 T_2] + \frac{sY}{ARL1} +$$

$$W + \frac{(a+bn)}{h}[\frac{1}{\lambda} - \tau + h(ARL2) + nE + \delta_1 T_1 + \delta_2 T_2]\,. \qquad (2.8)$$

The ratio of (2.8) to (2.3) is C, the expected cost per hour, or

$$C = \frac{\text{in-control} + \text{out-of-control} + \text{false} + \text{real} + \text{sampling}}{\text{cycle time}}$$

$$C = \frac{\frac{C_0}{\lambda} + C_1[-\tau + h(ARL2) + nE + \delta_1 T_1 + \delta_2 T_2] + \frac{sY}{ARL1}}{\frac{1}{\lambda} + \frac{(1-\delta_1)sT_0}{ARL1} - \tau + h(ARL2) + nE + T_1 + T_2}$$

$$+ \frac{W + \frac{(a+bn)}{h}[\frac{1}{\lambda} - \tau + h(ARL2) + nE + \delta_1 T_1 + \delta_2 T_2]\,]}{\frac{1}{\lambda} + \frac{(1-\delta_1)sT_0}{ARL1} - \tau + h(ARL2) + nE + T_1 + T_2} \qquad (2.9)$$

This completes the development of the cost model. All of the terms in equation (2.9) are in closed form except ARL1 and ARL2. Once we know the form of these, we can determine the variables which minimize (2.9). The next chapter explains the optimization procedure used to minimize (2.9) and how we calculate the average run lengths for the Shewhart, CUSUM and GMA charts.

III. Cost Model Solution

3.1 OVERVIEW

This chapter discusses the problem of minimizing the cost function, equation (2.9). We investigate the convexity of the function and present algorithms to find the minimum cost solution. We explain the computation of average run lengths for the Shewhart, GMA and CUSUM charts, all used in the cost model. A numerical example demonstrates the use of the algorithms.

3.2 CONVEXITY OF COST MODEL

We want to know that equation (2.9) is convex so that we can find the set of variables that yields the global minimum to our model. Applying first-order conditions that the gradient must be zero at a minimum or maximum, we determine the first partial derivatives of equation 2.9. To simplify the writing of the first derivatives, let DEN represent the denominator and NUM the numerator of equation (2.9). The first partial derivative of equation (2.9) will have the following form:

$$\frac{\partial C}{\partial x} = \frac{1}{DEN^2}\left[DEN\frac{\partial NUM}{\partial x} - NUM\frac{\partial DEN}{\partial x}\right] = 0, \qquad (3.1)$$

where x represents a decision variable. The sample size, n, is a discrete variable and the rate of change in equation (2.9) with respect to n would rigorously be determined using difference functions. However, for notational clarity and to demonstrate the concept of the effect of sample size on the cost function, we will assume for the moment that n is continuous and will take the partial derivative of the numerator and denominator with respect to n. The first partial derivatives of the numerator of equation (2.9) with respect to sample size, *n*, sampling interval, *h*, and sensitivity factor, *L*, are as follows:

$$\frac{\partial NUM}{\partial n} = \left[C_1 + \left(\frac{a+bn}{h}\right)\right]\left[h\frac{\partial ARL2}{\partial n} + E\right]$$

$$-\left[\frac{sY}{ARL1^2}\right]\left[\frac{\partial ARL1}{\partial n}\right] + \frac{b}{h}\left[\frac{1}{\lambda} - \tau + h(ARL2)\right]$$

$$+\frac{b}{h}\left[nE + \delta_1 T_1 + \delta_2 T_2\right] \qquad (3.2)$$

$$\frac{\partial NUM}{\partial h} = C_1\left[\frac{-\partial\tau}{\partial h} + ARL2\right] + \frac{Y}{ARL1}\left(\frac{\partial s}{\partial h}\right)$$

$$+\left[\frac{a+bn}{h^2}\right] \times h\left[\left(\frac{\partial\tau}{\partial h}\right) + ARL2\right]$$

$$-\left[\frac{a+bn}{h^2}\right] \times \left[\frac{1}{\lambda} - \tau + h(ARL2)\right]$$

$$-\left[\frac{a+bn}{h^2}\right] \times \left[nE + \delta_1 T_1 + \delta_2 T_2\right] \qquad (3.3)$$

$$\frac{\partial NUM}{\partial L} = -(2h)\left[\frac{\partial ARL2}{\partial L}\right] - \frac{sY}{(ARL1)^2}\left[\frac{\partial ARL1}{\partial L}\right] \qquad (3.4)$$

The first derivatives of the denominator of equation (2.9) with respect to sample size, *n*, sampling interval, *h*, and sensitivity, *L* are

$$\frac{\partial DEN}{\partial n} = \left[\frac{(1-\delta_1)sT_0}{ARL1}\right]\left[\frac{\partial ARL1}{\partial n}\right] + h\left[\frac{\partial ARL2}{\partial n}\right] + n \qquad (3.5)$$

$$\frac{\partial DEN}{\partial h} = \frac{(1-\delta_1)sT_0}{ARL1}\left(\frac{\partial s}{\partial h}\right) - \left(\frac{\partial \tau}{\partial h}\right) + ARL2 \tag{3.6}$$

$$\frac{\partial DEN}{\partial L} = \left[\frac{(1-\delta_1)sT_0}{ARL1^2}\right]\left[\frac{\partial ARL1}{\partial L}\right] + h\left[\frac{\partial ARL2}{\partial L}\right] \tag{3.7}$$

The average run lengths are functions of the sample size and sensitivity, so these equations include the partial derivatives of ARL1 and ARL2 with respect to *n* and *L*. The existence of these average-run-length terms means that the partial derivatives are not in closed form and we cannot compute the variables directly. Because the first derivatives are not in closed form, we also cannot use first and second order conditions to prove that equation (2.9) is convex and that a global minimum exists. Previous authors have found a minimum, which they assume to be the global minimum, of the cost function defined in (2.9) by using various search procedures. This assumes that the function is convex in the region of interest so that they can find at least a local minimum Duncan (1956) states that numerical studies of the cost model suggest that for reasonable values of the parameters, a local minimum does exist in the neighborhood of values of *n, h, L* used in the heuristic design of control charts.

We examined the convexity of the cost function for a restricted range of variable values. Figures 3.1 - 3.3 show how the cost per hour for a Shewhart chart changes as a function of one decision variable (*n, h, L*), while the other two variables are held constant. These graphs show that the function is convex in sampling interval, *h*, and sensitivity level, *L* . Figure 3.1 indicates that the function is not strictly convex in *n*, the sample size. We can approximate the function by a linear function, and we can find a locally optimal value of *n* in a narrow range of possible values. We can therefore assume that the function is convex over our range of *h* and *L* and that if we search over a narrow range of values for *n*, we will find a value of *n* which yields a local minimum of (2.9). It would be useful to know how the cost function behaves as *n* gets very large and as *L* increases in the negative and positive directions. The value of *n* influences the cost function in sampling costs, inspection time, and average run lengths. The cost of sampling and inspection time are increasing in n which will increase costs as n increases. The values for the average run lengths decrease as n increases. A lower value for ARL1 means more frequent false alarms and higher false alarm costs. A lower value for ARL2 means less time spent in the out-of-control state and fewer defects produced at the out-of-control rate. As long as the three increasing costs offset the one decreasing cost, the cost will continue to increase as n increases. As the sensitivity factor, L, increases, the average run lengths increase. The larger value for ARL1 means that we spend less on false alarms, while the larger value for ARL2 means we produce more defects at the out-of-control rate before we detect the shift. As long as the decrease in false-alarm costs offsets the increase in defect costs, the hourly cost will continue to increase as L increases. If L takes on negative values, there will be no effect on the cost graph because we compute the control limits using plus and minus L, which means that the graph will be symmetric about L=0.

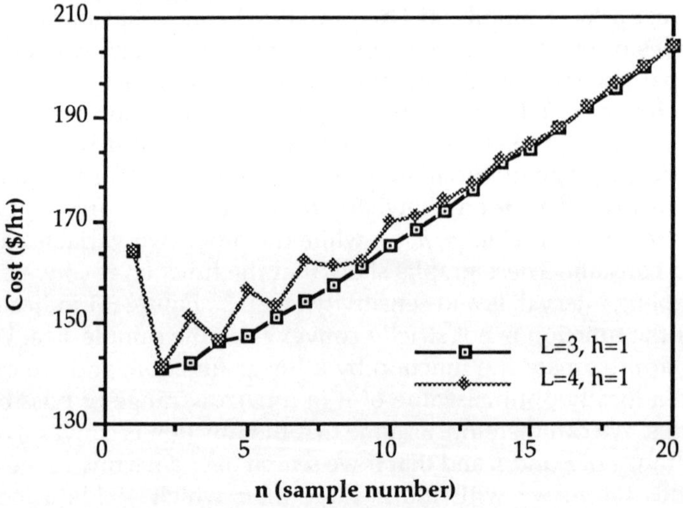

Figure 3.1 Cost vs. Sample Size

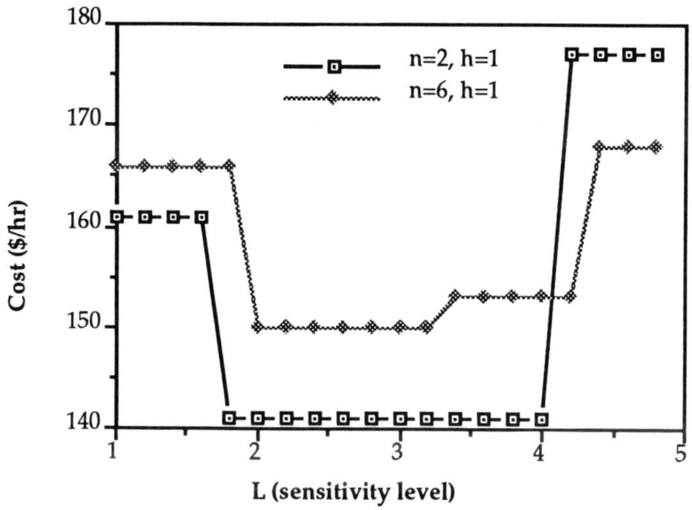

Figure 3.2 Cost vs. Sensitivity Level

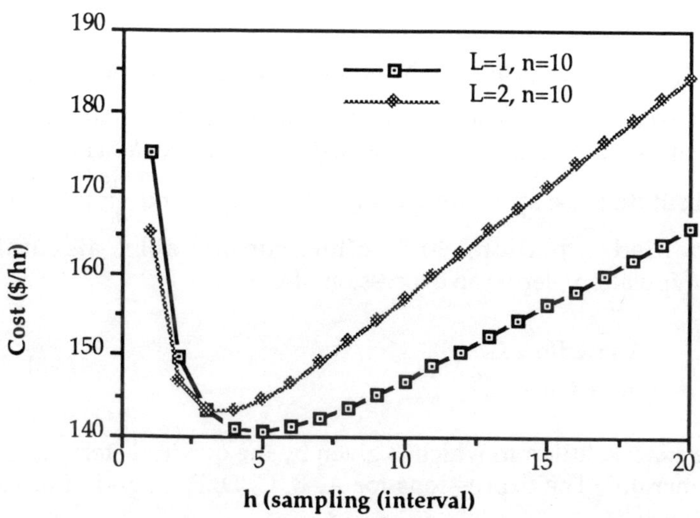

Figure 3.3 Cost vs. Sampling Interval

3.3 ALGORITHM

The prior investigations of the Duncan cost model have assumed that a local minimum, which is close to the global minimum, can be found by using a series of univariate search procedures. The performance of a search procedure improves if we find an initial value which is close to the optimal solution for the variable of interest . Lorenzen and Vance (1986) develop an approximation to the partial derivative of equation (2.9) with respect to *h* which allows them to compute an initial value for *h*. They expand the exponential terms in τ, s, $\frac{\partial \tau}{\partial h}$, and $\frac{\partial s}{\partial h}$ and ignore all terms containing powers of λ greater than 1. They justify this simplification because λ is small enough, relative to the other parameters, that in the approximation terms of λ^2 or greater become insignificant. This results in $\tau \approx .5h$, $s \approx \frac{1}{\lambda h}$, $\frac{\partial \tau}{\partial h} \approx .5$, and $\frac{\partial s}{\partial h} \approx \frac{1}{\lambda h^2}$. They state that equation (2.9) is numerically difficult to minimize because of the terms $\frac{C_0}{\lambda}$ in the numerator and $\frac{1}{\lambda}$ in the denominator. After multiplying the numerator by λ and factoring out C_0 to get a new form of the cost equation, they compute the first derivative with respect to *h*. They then substitute these approximate terms for τ, $\frac{\partial \tau}{\partial h}$, s, and $\frac{\partial s}{\partial h}$ into their modified equation, and after considerable algebraic manipulation, derive an expression of the form

$$\frac{Ah^2 + Bh + C}{(Fh^2 + Gh + H)^2} = 0, \qquad (3.8)$$

the exact solution to which is given by the quadratic term in the numerator. The expressions for A, B, C, D, F, G, and H are as follows:

$$A = KG - JH$$
$$B = 2(KF - IH)$$
$$C = JF - JG$$
$$F = \lambda(ARL2 - .5)$$
$$G = 1 + \lambda(nE + T_1 + T_2)$$
$$H = \frac{(1-\delta_1)\,T_0}{ARL1}$$
$$I = \frac{Y - C_0\,(1-\delta_1)T_0}{ARL1} + (a + bn)\,[1+\lambda(nE + \delta_1 T_1 + \delta_2 T_2)]$$
$$J = \lambda[(C_1 - C_0)(nE + T_1 + T_2) - C_1\,[(1-\delta_1)T_1 + (1-\delta_2)T_2] + W + (a+bn)(ARL2 - .5)]$$
$$K = \lambda[(C_1 - C_0)(ARL2 - .5)]$$

They then use the solution to the quadratic equation in the numerator of (3.8) as the starting value for h in the optimization algorithms. A starting value for n and L based on equation (3.1) is not as easy to obtain, and Lorenzen and Vance use heuristics to determine starting values. Montgomery (1980) suggests that we determine sample size largely by the magnitude of the shift Δ. For $\Delta \geq 2$, generally n is between 2 and 10. When Δ is between 1 and 2, n is generally between 10 and 20. For Δ between .5 and 1, n generally ranges from 20 to 40. Thus we may assume that an approximate starting value for n is $20/\Delta$. Because operating in the out-of-control state is more expensive than operating in control, a relatively small value of ARL2, the expected number of samples required to detect a real shift, is desirable. Chiu and Wetherhill (1974) suggest that ARL2 is generally about 1.2 for the Duncan model, and we can use this observation to determine an initial value for L given a value for n.

We do not know the limits of Lorenzen and Vance's search intervals. We used this algorithm in the initial version of our computer program and sought to reproduce Lorenzen and Vance's results to verify that our implementation of the algorithm was correct.[1]

Lorenzen and Vance present an example of a Shewhart p chart for controlling the proportion of defective items in a process. They use the starting value methods listed above and the algorithm flowcharted in Figure 3.4 in their work.

Our result was a lower cost solution than that published by Lorenzen and Vance, and we could duplicate their results only after we added an integrality constraint on the limits of the control chart. This additional constraint increased the sample size and the cost of the solution. Table 3.1 compares our solution to that of Lorenzen and Vance. Lorenzen and Vance's solution is similar to the approximate guidelines given above for initial values. They take a sample of 20 units every 2.88 hours. Their sensitivity factor, *L*, has a value of 3.336. This design has an average run length in-control of 33.5 samples and an average run length out-of-control of 1.48 samples, which is close to Chiu and Wetherhill's recommendation. Our design takes a sample of 1 item every .379 hours with a sensitivity factor of 1.05. We get longer average run lengths of 73.53 in-control and 8.85 out-of-control. The smaller but more frequent samples, combined with longer average run lengths, result in a 4.9 percent savings over the Lorenzen and Vance solution.

We also found that Newton's method would not converge in *h*. This presumably occurs because the cost function is very flat near the minimum, causing the algorithm to diverge. These findings led us to develop our own algorithm for minimizing equation (2.9). We performed a line search over *n* and *L* and used the golden section search method on *h*. We based the interval for our search on an analysis of Figures 3.1 to 3.3. These figures indicate ranges for the variables in which we can expect to find the minimum cost. The initial interval for *n* is from 1 to 5. The algorithm has a provision to increase this interval if the cost is still decreasing when the sample size reaches the upper limit of five. Based on our analysis of Figure 3.2 the minimum cost value for the sensitivity level, *L*, is between 1 and 4. We use a step size of .1 over this interval because the cost is relatively insensitive to small changes in *L*. Figure 3.3 shows that the interval for *h* should be from a value less than 1 to 20. We chose to use a slightly wider interval of .1 to 25 to ensure that we found the minimum cost value for *h*. The flowchart of our new algorithm for a c chart appears in Figure 3.5.

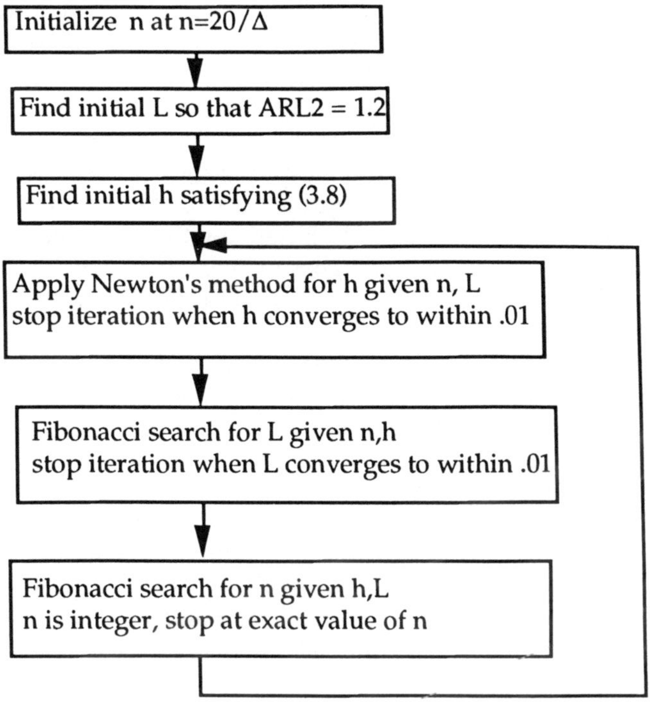

Figure 3.4 Lorenzen and Vance Flowchart

Table 3.1 Comparison of Results for P Chart Example

VARIABLE	Lorenzen / Vance	Matheson
n	20	1
L	3.336	1.05
h	2.88	.379
RESULTS		
ARL1	33.5	73.53
ARL2	1.48	8.85
COST/hr	237.21	225.44

The algorithm stops searching over *n* and *L* when it reaches an inflection point in the cost, i.e., COST(n | L,h) < COST(n+1 | L,h) and COST(L | n,h) < COST(L+1 | n,h). The golden section search for *h* stops when the cost converges to within .01.

For a GMA chart we use the same intervals for *n* , *L* and *h* and add an interval for the weighting factor, *r*. The weighting factor is between 0 and 1. Hunter (1986) states that based on econometric data r=.2 ± .1. Based on this observation, our interval for *r* starts at .1 and goes to .9 to ensure coverage of the most likely values for *r*. The flowchart of this algorithm appears in Figure 3.6.

For the GMA chart, the loop on *r* stops when the cost remains constant for three consecutive values of *r*. We developed this heuristic after we observed that the cost decreased as *r* increased, up to a point, and then the cost remained constant for the remaining values of *r* in the loop. We used the same stopping rules for *n* and *L* as in the c chart algorithm.

In the CUSUM algorithm, we added the scaling factor, *k*, and the decision interval, L_{cu}, while we deleted the sensitivity factor, *L*. We use the same intervals for *n* and *h* that we use in the c and GMA chart algorithms. The interval for *k* is .1 to 2. This interval was chosen because average run lengths increase very rapidly for values of *k* less than .1 and are on the order of 3 to 4 when $k > 2$. We found that large average run lengths did not produce minimum cost solutions and that we could avoid searching over small values of *k*. The interval for L_{cu} is .1 to f(k) where f(k) is the largest value of L_{cu} which the CUSUM average run length program uses for a particular value of *k*. We found that the cost did not change as a function of the average run length when the average run length was on the order of 100,000 samples or more. For a given value of *k* the average run length is increasing in L_{cu}. To reduce the solution time of the algorithm we truncate the range of L_{cu} to avoid calculating average run lengths greater than about 100,000 samples. We found this limit by analyzing tables of average run lengths in Vardeman and Ray (1985) and determining the appropriate value of *Lcu* for a given *k*. The solutions we found never approached this 100,000 sample limit on average run lengths. The flowchart for this algorithm appears in Figure 3.7.

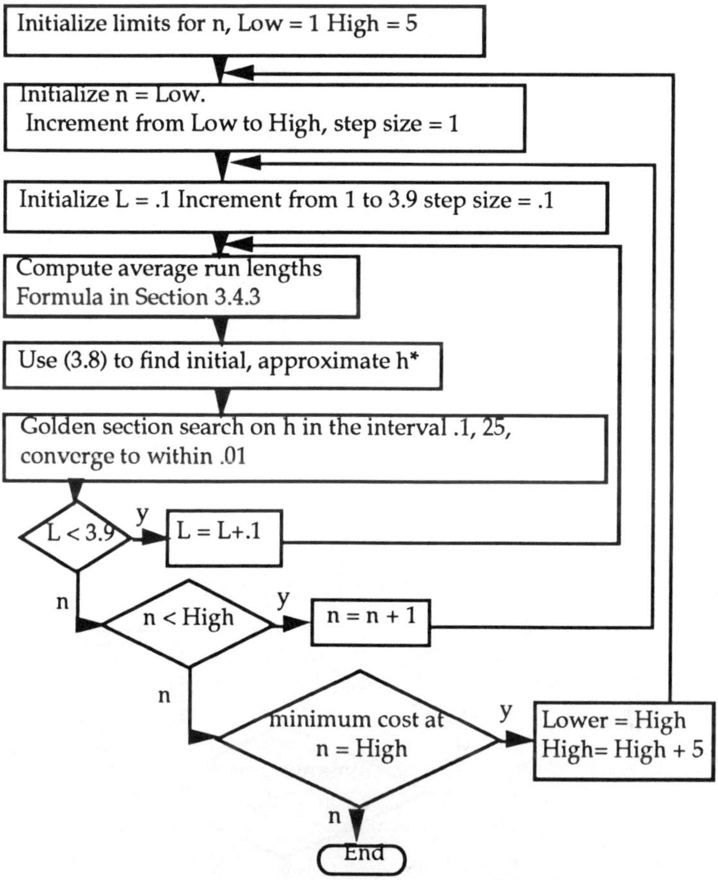

Figure 3.5 c Chart Flowchart

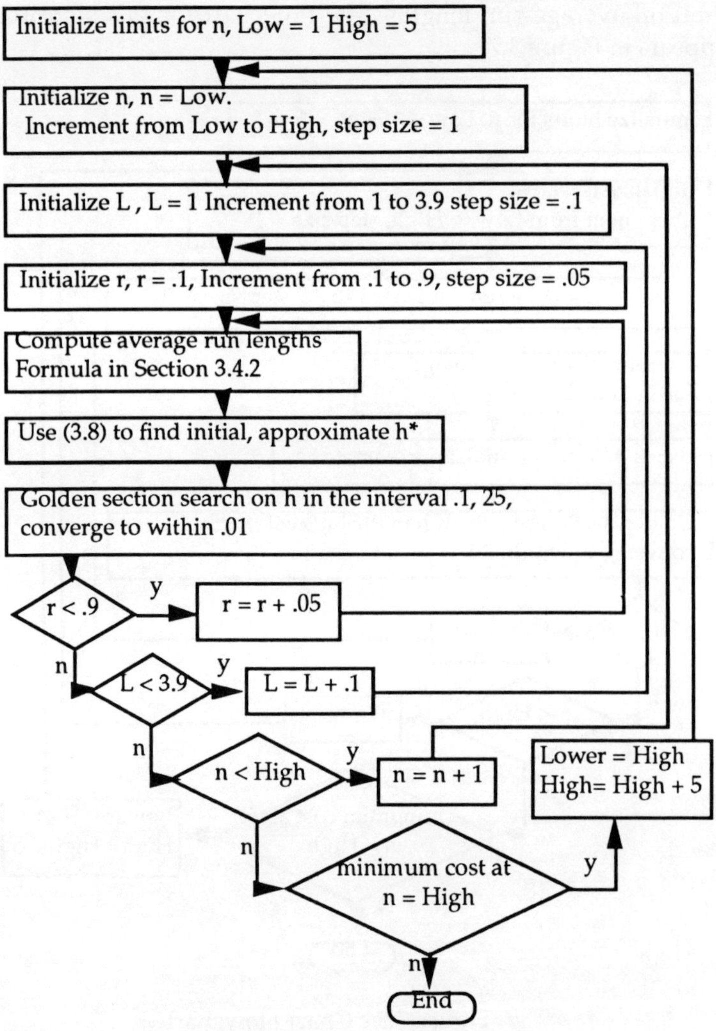

Figure 3.6 GMA Chart Flowchart

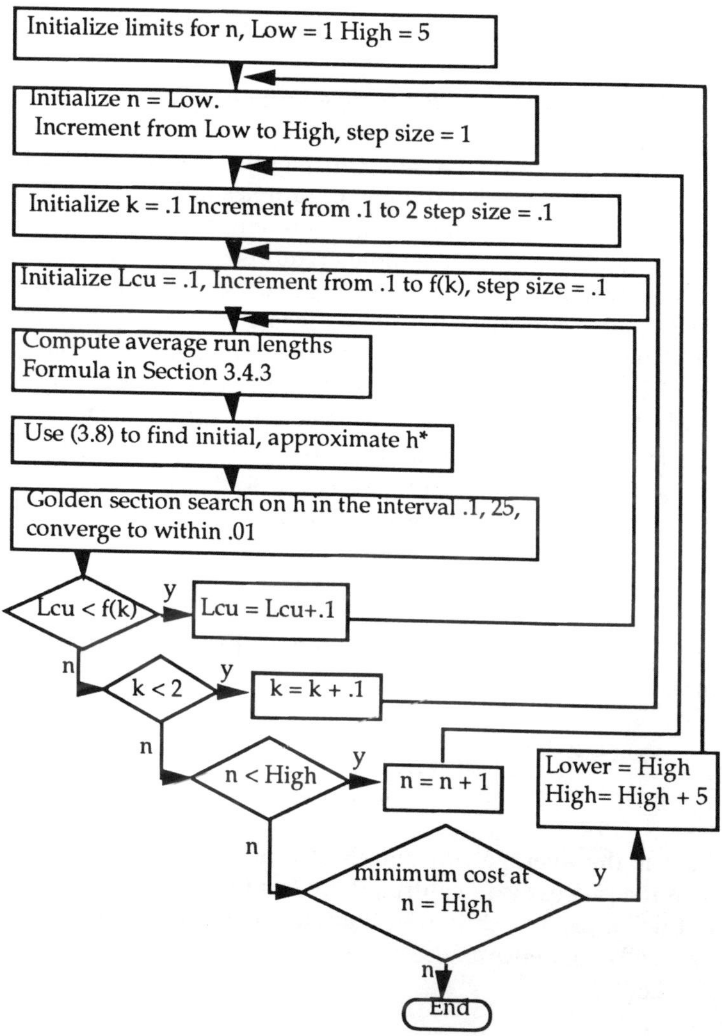

Figure 3.7 CUSUM Chart Flowchart

3.4 AVERAGE RUN LENGTH MODELS

3.4.1 Shewhart Chart Average Run Lengths

The concept of average run lengths was introduced in Section 1.2. The expected number of samples taken before a false alarm occurs is called the average run length in control, *ARL1*. The expected number of samples required to detect an actual out-of-control condition is the average run length out of control, *ARL2*. Each point plotted on a Shewhart control chart (Figure 2.1) is an independent Bernoulli trial of the in-control hypothesis. Therefore, the number of samples required to detect a shift is distributed geometric(1-p), where (1− p) is the probability of detection. For us to detect the shift on the kth sample, the first k–1 samples must have failed to detect the shift. The probability of first detecting the shift on the kth sample is

$$p\{\text{detect on kth sample}\} = p^{k-1}(1-p).$$

The expected number of samples analyzed before we detect the shift is

$$\sum_{k=1}^{\infty} k\, p^{k-1}(1-p) = \frac{1}{1-p},$$

which is the average run length of a Shewhart control chart. When the process is in control, there is a probability, α, that the chart will indicate a shift even though one has not occurred; that is, we will experience a false alarm.

Let

$$\alpha = p(\text{sample mean} < LCL \text{ or sample mean} > UCL \mid \text{mean} = \mu_0)$$

be the probability of a false alarm when the process is operating at the in-control mean μ_0. When the process goes out of control, the process mean shifts to a higher level

$$\mu_1 = \mu_0 + \Delta \sigma ,$$

where Δ is the magnitude of the shift in standard deviations. As we discussed in Chapter II, we are usually interested in detecting decreases in quality as indicated by higher defect rates. If the process has shifted, there is a probability, β, that the shift will go undetected, where

$$\beta = p(\text{ LCL} < \text{sample mean} < \text{UCL} \mid \text{mean} = \mu_1).$$

The average run length in control, ARL1, for a Shewhart chart is

$$\text{ARL1} = \frac{1}{\alpha}. \tag{3.9a}$$

The average run length out of control, ARL2, for a Shewhart chart is

$$\text{ARL2} = \frac{1}{(1-\beta)}. \tag{3.9b}$$

We assume that a Poisson process with in-control mean Λ_0 and out-of-control mean Λ_1, $\Lambda_0 < \Lambda_1$, generates the defects in our system. The Poisson probability function is

$$f(x,\Lambda_0) = \Lambda_0^{\ x} \ \frac{e^{-\Lambda_0}}{x!}.$$

We can compute the values of α and β by summing this function over the correct range of values for x. Define $\lceil y \rceil$ as the smallest integer greater than or equal to y and $\lfloor y \rfloor$ as the largest integer less than or equal to y. Because the number of defects is integer and the control limits are continuous, we assume that the process is in control as long as the number of defects, x, satisfies the relation $\lceil \text{LCL} \rceil \leq x \leq \lfloor \text{UCL} \rfloor$. We assume that the process is out of control when we have $x \leq \lfloor \text{LCL} \rfloor$ or $\lceil \text{UCL} \rceil \leq x$.

The values of α and β for a c chart are

$$\alpha = \sum_{i=\lceil ucl \rceil}^{\infty} (n\Lambda_0)^i \frac{e^{-n\Lambda_0}}{i!} + \sum_{i=0}^{\lfloor lcl \rfloor} (n\Lambda_0)^i \frac{e^{-n\Lambda_0}}{i!} \tag{3.10a}$$

$$\beta = \sum_{i=\lceil lcl \rceil}^{\lfloor ucl \rfloor} (n\Lambda_1)^i \frac{e^{-n\Lambda_1}}{i!} \tag{3.10b}$$

3.4.2 GMA Chart Average Run Lengths

The sample statistics of a GMA chart are a weighted sum of prior observations, so they are not independent. Brook and Evans (1973) developed a Markov chain approximation for the computation of average run lengths for dependent statistics. Gardiner (1987) extended their method to the calculation of average run lengths for GMA charts.

These calculations are based on the idea that the transition of the process from the in-control to the out-of-control state can be viewed as a Markov chain process. If Z_{t-1} is the value of the geometric moving average at time t-1, and X_t is the current sample statistic at time t, then we can define the probability that the new moving average, Z_t, is less than Z_{t-1} as $F(Z_{t-1} \mid X_t) = P(Z_t \leq Z_{t-1} \mid X_t)$. Alternatively, a stochastic process $\{Z_t\}$ is said to have the Markovian property if
$P\{Z_{t+1} = j \mid Z_0 = k_0, Z_1 = k_1 \ldots, Z_{t-1}k_{t-1}, Z_t = i\}$
$= P\{Z_{t+1} = j \mid Z_t = i\}$, for t = 0,1, . . . and every sequence i, j, k_0, k_1, . . . , k_{t-1}. This says that the transition probability from state i to state j depends only on the process being in state i at time t-1 and obtaining the appropriate value for the sample statistic at time t. This assumes that Z_t is a discrete random variable, but the geometric moving average is a continuous random variable. To approximate this continuous variable and use the discrete Markov approach, we divide the range of the moving average into discrete intervals.

We partition the range of the geometric moving average into M+1 states and define the states zero through M–1 as the in-control states and state M as the out-of-control or *absorbing* state. When the geometric moving average exceeds a control limit, the Markov chain has transitioned to the Mth state, and the manufacturing process stops so no further transitions can occur. Therefore, the process cannot proceed from the Mth state by itself; it stays there until the process is returned to the in-control state.

If we define p(i,j) to be the one-step transition probability from states i to j, then we can represent these probabilities by an (M+1 * M+1) transition matrix, **P** (Figure 3.8).

$$\mathbf{P} = \begin{vmatrix} p(0,0) & p(0,1) & p(0,2) & \dots & p(0,m-1) & p(0,m) \\ p(1,0) & p(1,1) & p(1,2) & \dots & p(1,m-1) & p(1,m) \\ p(2,0) & p(2,1) & p(2,2) & \dots & p(2,m-1) & p(2,m) \\ \cdot & \cdot & \cdot & \dots & \cdot & \cdot \\ \cdot & \cdot & \cdot & \dots & \cdot & \cdot \\ \cdot & \cdot & \cdot & \dots & \cdot & \cdot \\ p(m,0) & p(m,1) & p(m,2) & \dots & p(m,m-1) & p(m,m) \end{vmatrix}$$

Figure 3.8 Finite State Transition Matrix.

This matrix defines the possible states for the control scheme, and the actual value of M determines the accuracy of the approximation. Gardiner found that thirty states gave good results for the cases he investigated. He tested his method for small values of the process mean and shifts in quality. The actual discretization and calculation of the p(i,j) is based on the width of the control chart and the number of increments in the approximation. Recalling that the control limits of the GMA chart are

$$UCL = \mu_0 + L\,\sigma_{gma}$$
$$LCL = \mu_0 - L\,\sigma_{gma},$$

the width of the [UCL, LCL] interval is $2\ L\ \sigma_{gma}$. Letting W_S represent the "width" of the transition states of the Markov chain representation, then

$$W_S = \frac{2\ L\ \sigma_{gma}}{M}$$

when we use M increments in the approximation.

We determine the number of defects X_t required to move from state Z_{t-1} to Z_t by computing the difference of Z_t with respect to X_t, the current observation. The geometric moving average is

$$Z_t = rX_t + (1-r)\ Z_{t-1},$$

so the first difference is

$$\Delta\ Z_t = r\Delta X_t + \Delta\ (1-r)\ Z_{t-1},$$

or

$$\Delta\ Z_t = r\Delta X_t.$$

If $\Delta\ Z_t > W_S$, then a transition to a new state will occur. Thus, if $\Delta\ Z_t = r\Delta X_t > W_S$, then $\Delta X_t > \frac{W_S}{r}$ will cause a transition to a new state. This relates the change in X to the transition matrix and allows the calculation of the average run lengths based on the Markovian process.

To facilitate the calculation of the average run lengths, the **P** matrix is partitioned into four parts. This is shown in Figure 3.9, where **R** is an m*m matrix, **A** is an m*1 column matrix, **0** is a 1*m row matrix and **1** is a 1*1 identity matrix. The elements of the **A** vector represent the absorbing states and can be shown to be equal to (**I**–**R**) where **I** is an m*m identity matrix.

$$\mathbf{P} = \left[\begin{array}{c|c} \mathbf{R} & \mathbf{A} \\ \hline \mathbf{0} & \mathbf{1} \end{array}\right]$$

Figure 3.9 Partitioned Transition Matrix

Let N_i be the number of steps taken, starting from state i, to reach the absorbing state M for the first time and let $\mu_i^{(s)}$ be the sth factorial moment of N_i, where

$$\mu_i^{(s)} = E\{N_i^{(s)}\} = E\{N_i(N_i - 1) \ldots (N_i - s + 1)\}.$$

In matrix form this becomes

$$(\mathbf{I} - \mathbf{R})\ \mu^{(s)} = s\,\mathbf{R}\,\mu^{(s-1)} \tag{3.11}$$

When s = 1 we obtain the first moment (the mean) and (3.11) reduces to

$$(\mathbf{I} - \mathbf{R})\ \mu = \mathbf{1},$$

where the vector **1** is a vector of M ones. We can solve this series of equations via pivotal condensations, starting with the last diagonal element. Gardiner developed a computer program to compute the average run lengths for the GMA chart, and we used that program in this research.

3.4.3 CUSUM Chart Average Run Lengths

Section 2.2.2 explained the CUSUM chart with its scaling factor or target value k and sensitivity level L_{cu}. The CUSUM chart plots the cumulative sum of the deviations of the sample statistics, x, from the target value k. The mth cumulative sum is

$$S_m = \sum_{i=1}^{m} (x_i - k),$$

where x_i is the sample statistic of the ith sample. Figure 2.4 shows the hypothesis test regions of the CUSUM chart. An alternative form of the CUSUM considers that there are actually two tests being run simultaneously, one to detect increases in quality and one to detect decreases in quality. We discussed the nature of someone's interest in detecting an increase in quality in Chapter II. These revised cumulative sums are

$$S_i = \max\,[0\,, S_{i-1} + (X_i - k)],$$

which detects increases in the statistics x_i or shifts to the high side, and

$$T_i = \min\,[0\,, T_{i-1} + (X_i - k)]$$

which detects decreases in the statistics x_i or shifts to the low side. Here k is the same target value as before, and S_0 and T_0 are the initial values for the cumulative sum which is typically set at zero (Vardeman and Ray, 1985). The sensitivity level for the increasing sum is $+L_{cu}$, and for the decreasing sum, it is $-L_{cu}$. The average run length for the CUSUM chart is the expected number of samples taken before $S_i > L_{cu}$ or $T_i < -L_{cu}$. The average run length is a function of the starting value for the sum, S_0, and the probability that we will observe a particular value of x.

Define H(s) to be the average run length of a CUSUM chart intended to detect shifts to the high side starting from initial value s and let $f(\cdot)$ be the probability function of x_i. To motivate the equation for average run lengths, (developed by Page in his 1954 article), consider the value of the cumulative sum at the ith sample, S_i, the ith observed statistic X_i, and the expected number of samples taken before the sum increases from S_i to L_{cu}, which is $H(S_i)$. Getting a value of $S_i = 0$ requires that $X_i \leq k - S_{i-1}$, which yields

$$S_i = S_{i-1} + (X_i - k) = S_{i-1} + (k - S_{i-1}) - k \leq 0,$$

but, since $S_i = \max[0, S_{i-1} + (X_i - k)]$, $S_i = 0$. With S_i at zero, we expect to wait H(0) samples before reaching the limit L_{cu}. Getting a value of $S_i = 1$ requires that $X_i = 1 + k - S_{i-1}$, which yields

$$S_i = S_{i-1} + (X_i - k) = S_{i-1} + (1 + k - S_{i-1}) - k = 1,$$

and we expect to wait H(1) samples before reaching the limit L_{cu}. We can extend this analysis up to the case where $S_i = h$. Dropping the dependence on i, letting *s* replace S_i, and taking the expectation over all possible values of *s*, we get

$$H(s) = 1 + H(0)\,F(k-s) + \int_0^{L_{cu}} H(y)\,f(y+k-s)\,dy, \qquad (3.12)$$

The number 1 arises because at least one sample is required to increase the sum from $s<L_{cu}$ to $s>L_{cu}$. A similar expression can be developed for the average run length to the low side, L(t).

Equation (3.12) is a general expression for the CUSUM average run length and Goel and Wu (1971) and Lucas and Crosier (1982) have evaluated it for normal variables. Vardeman and Ray (1985) developed a solution method and computer program which solves this integral equation for Poisson processes. They allowed us to use a copy of their code in our research.

3.5 NUMERICAL EXAMPLE

Before comparing the overall performance of the cost model for the three charts, we present a numerical example to demonstrate the inputs and outputs of the cost model.

We model a process which creates defects in its output at the rate of .1 defects per unit when operating in control. This defect rate increases to .416 defects per unit when a shift occurs.

We can make several assumptions about the choice of out-of-control defect rate. We can assume that there is only one out-of-control defect rate, which is the standard assumption of the earlier works on single state systems. Or we can assume that there are several rates depending on what happens to the machine, the multiple-state system. If there are several possible out-of-control defect rates, then we can use a conservative stopping rule and stop when the control chart indicates that we have exceeded the smallest of these rates. This hypothetical process remains in control an average of $\frac{1}{\lambda} = 100$ hours. A defect costs \$10.00 to repair. Sampling costs are \$1.00 to set up the sample and \$4.00 per unit. Analysis of a sample takes 5 minutes. Searching for a false alarm takes 5 minutes and costs \$200. Searching for the cause of a real shift takes 6 minutes and process repair takes 30 minutes. The cost of finding and repairing a real cause is \$500.00. The process is stopped whenever a shift is detected and the system produces 100 units per hour. We have listed these parameters in Table 3.2.

Optimization programs use these parameters for each chart to determine the set of variables that minimizes the hourly cost of quality. Table 3.3 shows the optimal chart designs for the c, GMA and CUSUM charts.

The design for the c chart takes a random sample of four units every 2.54 hours. The upper control limit is

$$UCL = n\,\Lambda_0 + L\,\sqrt{n\,\Lambda_0} = 4^*.1 + 2.5\,\sqrt{4^*.1} = 1.98,$$

which means that a shift is indicated if more than 1 defect is found in the sample. The lower control limit is

$$LCL = \max\,(0, n\,\Lambda_0 - L\,\sqrt{n\,\Lambda_0} = 4{*}.1 - 2.5\,\sqrt{4{*}.1}\,) = 0.$$

Table 3.2 Parameter Levels for Numerical Example

Times (hrs)		
Between shifts	$\frac{1}{\lambda}$	100
Sample/chart 1 item	E	.083
False search time	T_0	.083
Real search time	T_1	.1
Real repair time	T_2	.5
Costs ($)		
Repair of a defect	D	10
False alarm	Y	200
Real alarm	W	500
Fixed sampling	a	1
Variable sampling	b	4
Miscellaneous		
Magnitude of shift (std dev)	Δ	1
Defects per unit		
in-control	Λ_0	.1
Production		
during search	δ_1	0
during repairs	δ_2	0
Production rate (units / hr)	Θ	100

Section 2.2.1 describes this modification of the control limits, and it means that a sample with zero defects is still in-control. We can calculate the average run length in control , ARL1, using equations (3.9 a) and (3.10a). Since a sample value equal to the LCL of zero does not indicate that the process is out of control, we can modify equations (3.10a) and (3.10b) to reflect this change.

Table 3.3 Variables for C, GMA and CUSUM Charts

VARIABLE	C	GMA	CUSUM
n	4	1	1
L	2.5	2.2	na
r	na	0.35	na
k	na	na	0.2
L_{cu}	na	na	0.2
h	2.54	0.76	0.73
RESULTS			
ARL1	16.25	43.29	58.07
ARL2	2.02	6.09	5.93
COST($/hr)	127.83	129.49	127.50
na = not for this chart			

$$\alpha = \sum_{i=\lceil UCL \rceil}^{\infty} (n\Lambda_0)^i \frac{e^{-n\Lambda_0}}{i!}$$

$$= \sum_{i=2}^{\infty} .4 \frac{e^{-.4}}{i!}$$

$$= .062$$

$$ARL1 = \frac{1}{\alpha}$$

$$= \frac{1}{.062}$$

$$= 16.25$$

We calculate the average run length out-of-control, ARL2, using equations (3.9b) and (3.10b)

$$\beta = \sum_{i=\lceil LCL \rceil}^{\lfloor UCL \rfloor} (n\Lambda_1)^i \frac{e^{-n\Lambda_1}}{i!}$$

$$= \sum_{i=0}^{0} 1.664^i \frac{e^{-1.664}}{i!}$$

$$=.5045$$

$$ARL2 = \frac{1}{1-\beta}$$

$$= \frac{1}{1-.5045}$$

$$= 2.02$$

ARL1 means that a false alarm occurs, on average, every 16.25 samples, which is equivalent to an average of h*ARL1 = 2.54 *16.25 = 41.27 hours between false alarms. ARL2 means that the chart indicates a real shift after an average of only 2.02 samples. Since there are 2.54 hours between samples, it takes an average of h*ARL2 = 2.54 *2.02 =5.13 hours to detect the shift.

The design for the GMA chart samples a single unit every .76 hours. The upper control limit is

$$UCL = \Lambda_0 + L\,\sigma_{gma}.$$

This is $0.4203 = .1 + 2.2 * \sqrt{.1}\sqrt{\frac{.35}{2-.35}}$, and a shift is indicated if the moving average exceeds .4203. If a manufacturer uses this GMA design, he would expect a false alarm every 43.29 samples, which is the same as

h*ARL1 = .76 * 43.29 = 32.9 hours.

A real alarm occurs an average of 6.09 samples after the shift occurs. With a sampling interval of h=.76 hours, it takes an average of 6.09 * .76 = 4.62 hours to detect the shift.

The c chart has smaller average run lengths and longer sampling intervals than the GMA chart. In this case, the GMA chart detects false and real shifts more quickly than the c chart, but it costs $1.66 more per hour to operate the GMA chart than the c chart. This cost penalty is due to the larger number of false alarms per hour. This relationship between average run lengths and sampling interval indicates that the frequency of alarms is not an accurate indicator of the cost of operating a control chart.

The design for the CUSUM chart takes a sample of one unit every .73 hours. The values plotted on the chart are the cumulative sums of deviations from the target, k=.2, where the ith cumulative sum is

$$S_i = \sum_{j=0}^{i} (X_j - k).$$

If the cumulative sum exceeds the value of L_{cu}= .2, then the chart indicates a shift, which means one defect in the sample means the process is out-of-control. This chart requires an average of 58.07 samples before a false alarm occurs, which appears to be more stable than the c chart, but it also requires 5.93 samples, on average, to detect a real shift. Like the case of the GMA versus c chart, the CUSUM and c charts have different sampling intervals and this comparison of ARL2 values is misleading. A more meaningful non-economic comparison, might be the number of hours which elapse before detection. For the c chart this is 5.13 hours, while for the CUSUM chart, it is an average of 4.33 hours. This lends support to the notion that purely statistical comparisons of performance are not very useful and that economic design can give a more complete picture of chart performance.

The costs of these designs are within a few dollars of each other, but they are only hourly costs. Compared on a yearly basis these differences appear more significant. The $.33 savings of the

CUSUM over the c chart means an annual savings of $2059 for around the clock manufacturing. The GMA design would cost $12,418 more than the CUSUM design for the same process. This savings tends to justify the use of optimally designed control charts.

3.6 COMPARISON OF COST PERFORMANCE

One goal of this research was to compare the performance of the control charts. We were particularly interested in the cost of the different control schemes as process quality increased. To improve the value of our analysis of the model's sensitivity we wanted to find those variables which had the largest effect on the cost of quality. Since the cost model was virtually identical to the one proposed by Duncan (1956), we looked at prior investigations of the model's sensitivity. Panagos, Heikes and Montgomery (1985) performed a complete sensitivity analysis of the Duncan model for continuous and discontinuous production. They found that the cost of quality control is significantly affected by three factors: the mean time in control, the increase in the hourly cost of defects when the process goes out of control, and the time of search and repair for real shifts. These three factors relate to these parameters.

Mean time in control	$\frac{1}{\lambda}$
Increase in hourly cost of defects	$\Delta, \Lambda_0, D, \Theta$
Cost of search and repair for real shift	T_1, T_2

Using these parameters, we computed the increase in the hourly cost of defects as

$$C_1 - C_0 = (\Lambda_0 + \Delta\sqrt{\Lambda_0})D\,\Theta - \Lambda_0\,D\,\Theta = \Delta\sqrt{\Lambda_0}\,D\,\Theta.$$

All of these parameters are described in Table 3.4. We designed our analysis around these seven significant parameters so that we were only studying the effects of the significant factors. To provide a broad range of hypothetical processes, we set the seven parameters to high and low levels as would be done in a traditional designed experiment. The "design" was a 2^{7-1} fractional factorial using 64 combinations of parameters. In the design the seventh parameter is set to high or low levels based on the levels of the other six parameters. We used this method to provide a framework for developing the different parameter sets. The actual levels of the parameters we used are in Table 3.4.

Three factors indicate the quality of a process: how long it remains in control, $\frac{1}{\lambda}$, the number of defects per unit, Λ_0, and the magnitude of the shift, Δ. To simulate three quality levels at which to test the performance of each chart, we created ranges for the values of these three key parameters. We set the rest of the parameters at the same values for all three levels. This allowed us to compare the control charts solely on changes in quality. We repeated the 2^{7-1} fractional factorial system at each of these three quality settings.

To investigate the effect of continuity of production on the cost of quality, we also set the values of δ_1 and δ_2 at their four possible (0,1) combinations. This meant that a total of 64(levels) x 3(quality settings) x 4(continuity settings) x 3(different control charts) = 2304 computer runs of the cost model were made.

The overall results of this study are that the c chart is less expensive at the low-quality level, and the CUSUM chart is less expensive at the medium- and high-quality levels. The GMA chart occasionally matches the performance of the c and CUSUM chart but typically it is more expensive. We discuss the results of our analysis at each quality level in the remainder of this section. We find that the settings of the continuity of production variables do not alter the selection of the minimum cost control chart, although the values of δ_1 and δ_2 cause significant differences in cost for a specific chart. Section 3.8 discusses the implications of the continuity variables.

We begin our specific discussions with the low-quality system characterized by an average of 0.2 defects per unit of product and an average of 75 hours in control. The average magnitude of the shift in process quality is 1.5 standard deviations. Table 3.4 shows the average hourly cost by chart and quality level. While the CUSUM chart is less expensive than the c chart in 125 of 256 cases, the average savings is only 0.7 percent. The average savings of the CUSUM over the c chart are overstated because there are sixteen cases when the in-control hourly cost of defects is at high levels, where the CUSUM chart is approximately 5 percent less expensive than the c chart. If we ignore these sixteen large savings, the CUSUM actually costs more than the c chart, which means that at relatively low quality levels, the c chart would be the least expensive chart.

The medium-quality system has an average of 0.02 defects per unit of product and remains in control an average of 266 hours. The average magnitude of the shift in process quality is 0.15 standard deviations. The CUSUM chart is always the least expensive chart to use for the medium-quality system. The cost savings for the CUSUM over the c chart is 4.6 percent, an average dollar difference of $4.44 per hour.

The high-quality system has an average of 0.002 defects per unit of product and remains in control an average of 750 hours. The average magnitude of the shift in process quality is 0.015 standard deviations. The CUSUM chart is the least expensive chart, although the c chart has the same cost as the CUSUM chart in sixteen cases. The average cost savings of the CUSUM chart over the c chart is 9.37 percent, an average difference of $1.00 per hour.

This analysis indicates that the CUSUM chart can offer significant savings over the c and GMA charts if computer programs are available to design the charts. There is, however, a stronger observation to draw from this analysis. Table 3.5 shows the possible savings of improving process quality. At the current cost parameters, a company could save $1.67 million per year if it improved from a low-quality to a medium-quality level. This value is for one eight hour shift per day. That much money would support a large investment in manufacturing technology and quality control. While there are decreasing marginal returns for improving from medium quality to high quality, there is still a potential annual savings of $172,178.

Table 3.4 Parameter Levels for Cost Comparisons

PARAMETERS		QUALITY LEVELS		
Times (hrs)		LOW	MEDIUM	HIGH
Between shifts	$\frac{1}{\lambda}$	50 / 100	200 / 333	500 / 1000
Sample/chart 1 item	E	.083	.083	.083
False search time	T_0	.083	.083	.0083
Real search time	T_1	.1 / .3	.1 / .3	.1 / .3
Real repair time	T_2	.5 / 1	.5 / 1	.5 / 1
Costs ($)				
Repair of a defect	D	10 / 100	10 / 100	10 / 100
False alarm	Y	200	200	200
Real alarm	W	500	500	500
Fixed sampling	a	1	1	1
Variable sampling	b	4	4	4
Miscellaneous				
Magnitude of shift	Δ	1 / 2	.1 / .2	.01 / .02
Defects per unit in-control	Λ_0	.1 / .3	.01 / .03	.001 / .003
Production during search	δ_1	0 / 1	0 / 1	0 / 1
during repairs	δ_2	0 / 1	0 / 1	0 / 1
Production rate (units/hr)	Θ	50 / 100	50 / 100	50 / 100
x / y indicates two levels for bi-level variables				

Table 3.5 Cost Comparisons and Savings

CHART	AVERAGE HOURLY COSTS ($)		
	LOW	MEDIUM	HIGH
C	895.14	96.90	10.68
CUSUM	888.13	92.46	9.68
GMA	906.31	98.19	10.68
n = 256			
COST SAVINGS FROM QUALITY IMPROVEMENT ($)			
	LOW - MED	LOW - HI	MED - HI
HOURLY	796	878	83
ANNUAL	1,654,985	1,827,163	172,178

3.7 SENSITIVITY OF COST MODEL TO PARAMETER ESTIMATES

Saniga and Shirland (1977) and Chiu and Wetherhill (1974) report that very few practitioners have implemented economic models for the design of control charts. One reason for this is the difficulty of estimating costs and other model parameters. Managers are reluctant to use a model that requires them to estimate future costs which are hard to define. We investigated the sensitivity of the optimal solution to variation in the cost per defect. We increased the cost per defect by 5 and 10 percent and determined the optimum variables for the c chart at low quality levels. We also computed the cost of quality for a system based on the heuristic design of sample size $n = 20$, sensitivity level $L = 3$, sampling interval $h = 2$ hrs.

Table 3.6 shows the results of our sensitivity studies. The cost of the optimal solution increased by slightly less than the increase in the cost per defect. The heuristic solution was considerably more expensive than the optimal solution. On average, the heuristic solution was $111.24 (17.55 percent) per hour more expensive than the exact solution. This shows that the heuristic designs are significantly more expensive than the optimal designs.

Table 3.6 Effect of Difference in Defect Cost Estimates in Average Dollars per Hour

	5% error	10% error	Heuristic
AVG $/HR	896.44	934.74	969.36
Difference ($/hr)			
Max	123.88	247.76	620.88
Min	2.9	5.8	26.92
Average	38.31	76.61	111.24
% diff			
Max	4.83	9.67	45.2
Min	3.77	7.52	4.34
Average	4.44	8.88	17.55
AVG. EXACT = $858.12/hr			

Effect on Decision Variables

	n	L	h (5%)	h (10%)
Difference				
max	3	.6	-1.014	-1.071
average	1.8	.34	.006	-.024
avg(%)			1.4	1.5
# cases	10	10	56	56

The impact of estimation errors on the optimal design is another important sensitivity measure. If the design is robust to errors in cost estimates, then we will still have a nearly optimal design despite the errors. For the 64 cases we study, the optimal values of n and L change in only 10 cases. This suggests that the sample size and sensitivity interval, which are treated as discrete variables in the optimization algorithms, are not very sensitive to errors in cost estimates. There is no obvious reason different values for n and L occur for the ten cases. The sampling interval

h is a continuous variable, and its optimum value will change as the cost per defect changes. The optimal value for h changes in 56 of the 64 cases we investigated. On average, there is only a 1.4 percent change in the optimal value of h. However, there are 12 cases where there is more than 10 percent change in h. The conclusion of this sensitivity analysis is that errors in estimating the cost of defects have little effect on sample size and sensitivity level. The optimal value of the sampling interval will change, but on average, it will change only slightly. This analysis also supports the use of economic analysis rather than using the standard heuristic designs.

3.8 EFFECT OF PROCESS SHUTDOWN

This section discusses the significance of the decision to stop the process during the search and repair phases of the quality cycle (Figure 2.8). The continuity variables

$$\delta_1 = \begin{cases} 1 \text{ if production continues during search} \\ 0 \text{ otherwise,} \end{cases}$$

$$\delta_2 = \begin{cases} 1 \text{ if production continues during repair} \\ 0 \text{ otherwise,} \end{cases}$$

allow the application of the cost model to four possible production scenarios. They also allow us to study the effect of different management policies. If the process is shut down when the chart indicates a shift then the process spends less time in the out- of-control state. But, if the process is torn down to search for the assignable cause, then a setup cost would be incurred, raising the costs of search and repair. Alternatively, if the process does not stop, there is no tear-down, but the process spends more time operating in the out-of-control state which increases the number of defects needing repair. The cost model allows us to compare the cost of implementing these two policies.

To measure the impact of the continuity variables on the costs of our example systems, we ran the optimization programs at all four combinations of δ_1 and δ_2. Table 3.6 shows the average cost for each chart as a function of δ_1 and δ_2, based on the data in Table 3.4. The figures in Table 3.7 indicate that the choices of production shutdown policy have a significant effect on the cost of quality for low-quality systems, but the cost difference is insignificant in a high-quality system. The data in Table 3.7 also show the cost penalty of making the wrong choice for δ_1 and δ_2. For example, at low quality, the c chart can cost $75.68 or 8.8 percent more per hour if production is not stopped during search and repair. These examples indicate that the continuity variables can increase the usefulness of the cost model and allow it to extend to various types of design studies.

3.9 SUMMARY

This chapter investigated the convexity of the cost model (2.9), described the algorithms used to find the minimum cost variables, explained the calculation of average run lengths, and presented numerical results and findings.

While we cannot prove that equation (2.9) is convex, Figures 3.1 to 3.3 indicate that the function is probably convex in sampling interval, h, and sensitivity level, L. The function does not appear to be strictly convex in n, the sample size.

We show in Section 3.3 that we obtain a lower cost solution than that published by Lorenzen and Vance and develop new algorithms for computing minimum cost solutions. These algorithms use grid searches and the golden section method. Next we explain the calculation of average run lengths for the c, GMA and CUSUM charts.

Section 3.5 describes a numerical example and applies the cost model to the three control charts we are studying. Table 3.3 shows the results we obtained from the algorithms. Section 3.6 compares the performance of the three charts over a range of quality levels and parameter settings. We find that the c chart is generally more effective at the low-quality level, which has large shifts in quality. The CUSUM chart is more effective at the medium- and high-quality levels which have smaller shifts in quality.

Table 3.7 Table of Costs for Continuity Variables ($/hr)

CONTINUITY		LOW QUALITY		
δ_1	δ_2	C	GMA	CUSUM
0	0	858.13	865.65	839.95
1	0	895.46	902.09	901.46
0	1	893.60	917.03	871.59
1	1	933.81	940.48	939.51

CONTINUITY		MEDIUM QUALITY		
δ_1	δ_2	C	GMA	CUSUM
0	0	96.51	97.83	92.04
1	0	96.82	98.11	92.41
0	1	96.98	98.28	92.51
1	1	97.28	98.56	92.88

CONTINUITY		HIGH QUALITY		
δ_1	δ_2	C	GMA	CUSUM
0	0	10.68	10.68	9.74
1	0	10.68	10.68	9.75
0	1	10.69	10.69	9.49
1	1	10.69	10.69	9.75

A sensitivity analysis in Section 3.7 shows that errors in cost parameter estimation have little effect on the value of the control chart variables. We show the heuristic solution to be significantly more expensive than the optimal designs. We also show how the algorithms can assess the effectiveness of different production control policies in Section 3.8

In summary, this chapter shows the effectiveness of the cost model and algorithms which we developed. The algorithms can determine the minimum cost control chart and the variables for that chart. The algorithms can also compare the cost of alternative process shutdown policies.

NOTES

1. It was difficult to obtain any details from Lorenzen and Vance other that those in their paper. We wrote to the authors, requesting additional information about the algorithm and its implementation, but did not receive any response. Lorenzen and Vance are employees of General Motors which uses this algorithm in its quality control programs.

IV. Statistical Quality Control And Process Design

4.1 INTRODUCTION

This chapter presents an extension of the economic design of control charts to the design of production systems. We determine the optimal control chart variables and amount of buffer inventory to hold between a process and subsequent assembly operation.

4.2 DESCRIPTION OF PROBLEM

The Just-In-Time and Kanban production philosophies seek to minimize inventory levels throughout a factory (Schonberger, 1982). Eliminating buffer inventory between stages is one way to achieve this. This is done to reduce inventory holding costs and expose quality problems which can be obscured by inventories. But this "zero inventory" policy might not be optimal if it is less expensive to inspect and repair components before they are assembled into a product. For example, inspection and repair of an electronic device will be less expensive if it is done before the device is encapsulated into the product. If we assume that the process experiences shifts in quality, then statistical quality control could be used to monitor process quality and indicate when the defect rate shifts to an unacceptable level. But statistical quality control, with its periodic inspection, allows the process to operate at the out-of-control defect rate for a period of time before the shift is detected. If we want to inspect and repair the non-conforming units before assembly, then we need an inventory buffer to hold the items produced while the process is at the out-of-control level. The economic model of Chapter II contains defect and inspection parameters which can be extended to determine the optimal control chart variables and the size of the inventory buffer.

4.3 RELATIONSHIP TO BUFFERS IN PRODUCTION LINES

Our study of buffer inventories uses buffers to minimize the cost of quality for a single machine. We do not consider the effect of a shift in product quality on the production rate of an entire production line. The impact of inventory buffers on the performance of production lines has been studied by several authors, including Buzacott and Hanifin (1978) and Okamura and Yamashina (1979). They assume the production line consists of a series of unreliable machines which may stop working or fail due to operational or time-related causes. These machines always produce acceptable product when they are operating, so they need not consider statistical quality control. Observation of three stations, *k-1, k, k+1,* in a production line shows the usefulness of buffers. If station *k* fails, then station *k-1* might have to cease production due to a lack of storage space. Station *k-1* is said to be *forced down* or *blocked* by station *k*. Should station *k+1* run out of parts from station *k*, station *k+1* might be forced down or *starved* . The addition of buffer inventories between stages *decouples* the stages. Station *k-1* can produce until the buffer it feeds becomes full, and station *k+1* can operate until the buffer it draws from becomes empty. These buffers increase the time available to repair station *k* before its failure affects the rest of the line.

These authors have sought to optimize some performance measure, such as line throughput, subject to a constraint on buffer cost. They find that the use of buffer inventories can improve line throughput.

Our study of buffers involves one machine and the cost associated with using non-conforming components in a subsequent assembly stage. We seek to reduce the cost of quality through the use of buffers. Since we use the JIT policy of stopping the entire production line when a quality problem occurs, we do not need buffer inventories to prevent blocking and starving of stations.

The quality control and production line buffer problems can be integrated into a larger model for traditional production lines. This model would include the economic design of quality control charts and the use of buffers to decouple production and minimize the cost of quality. A complete production system could be designed by combining the quality and buffer models with the allocation of inspection model.

4.4 REVISION OF COST MODEL

Figure 4.1 depicts a production system which produces items at the rate of Θ units per hour. The items are joined with the product after they pass through a buffer which holds B units.

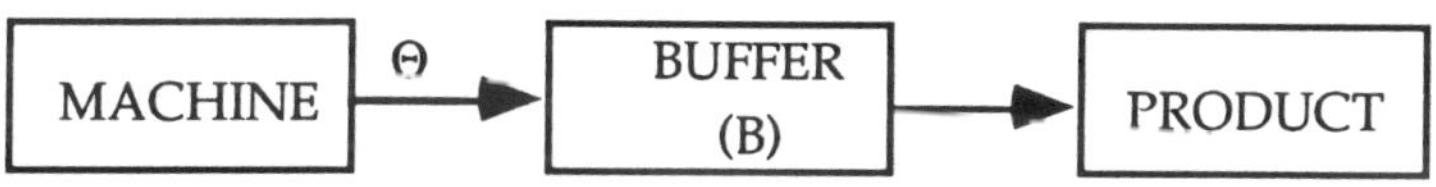

Figure 4.1 Diagram of Buffer System.

The size of the buffer, *B*, is an additional decision variable in our revised cost model. We assume that we are operating a balanced production line with each stage operating at the same production rate and the buffer is always full of units. We also assume that the process and the rest of the production line stops as soon as a shift in quality is indicated on the control chart and all of the units produced after the shift are inspected. To reduce the cost of inspection we do not inspect units in the buffer until we know that a real shift has occurred. We are assuming that there is sufficient inspection capacity to inspect all of the items produced after a shift during the repair period. This means that the cycle time is independent of the size of the buffer and the quantity produced after the shift.

4.4.1 Distribution Function of Quantity Produced After the Shift

The number of units in the buffer is a function of the number of items produced during the time the process is out-of-control. Let Q denote the quantity produced after the shift where

$Q = \Theta$ (time until shift is detected).

The time until we detect the shift consists of three increments (see Figure 4.2):

1. The random amount of time after the shift until we take the next sample, R, $0 \le R \le h$.

2. The time until we take a sample which contains enough defects to allow us to detect the shift, $(M\text{-}1)h$, where M is the random, integral number of samples and h is the predetermined time between samples.

3. The time to analyze and chart the results, nE, where n is a known sample size and E is a constant, per unit inspection time.

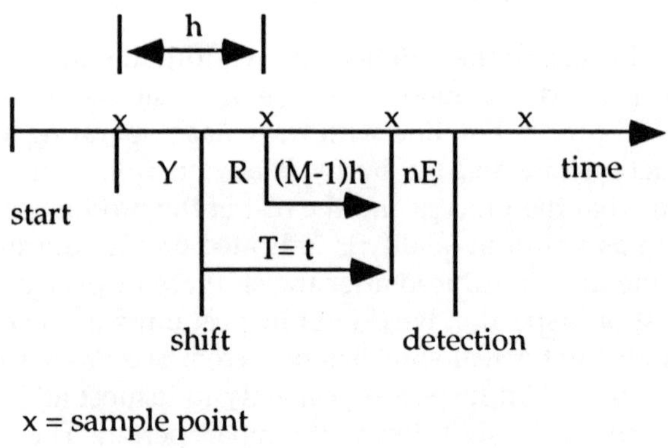

Figure 4.2 Diagram of Time Until Detection

Letting T represent the random components of the out-of-control time, we have

$$T = R + (M-1)h,$$

and the entire out-of-control time is T + nE. Therefore, the quantity produced between the shift and detection is

$$Q = (T+nE)\Theta.$$

The quantity Q is a random variable whose cumulative distribution we define as

$$\begin{aligned} FQ(q) &= p(Q \leq q) \\ &= p((T+nE)\Theta \leq q) \\ &= p(T \leq \frac{q}{\Theta} - nE) \end{aligned} \tag{4.1}$$

Equation (4.1) shows that the distribution of Q can be derived from the distribution of T through the transformation Q = (T+nE)Θ. Since T = R + (M-1)h we will develop the distributions for M and R before we derive the distributions of T and ultimately Q.

To develop the distribution for *M*, the number of samples taken after the shift, we start with the time between the shift and the next sample, the random variable R. Since

$$R = t - (M-1)h \text{ and}$$

$$0 \leq R \leq h \tag{4.2}$$

replacing R by (t-(M-1)h) in equation (4.2) yields the inequality,

$$\begin{aligned} &0 \leq t-(M-1)h \leq h \\ &-t \leq -(M-1)h \leq h-t \\ &\frac{t-h}{h} \leq M-1 \leq \frac{t}{h} \end{aligned}$$

$$\frac{t-h}{h}+1 \le M \le \frac{t}{h}+1. \qquad (4.3)$$

Since the number of samples is an integer, equation (4.3) becomes

$$\lceil t/h \rceil \le M \le \lceil t/h \rceil + 1, \text{which is equivalent to}$$

$$M = \lceil t/h \rceil + 1,$$

where $\lceil x \rceil$ is the smallest integer greater than x and $\lfloor x \rfloor$ is the largest integer less than x. Section 3.4.1 explained that since each point plotted on a Shewhart control chart is an independent Bernoulli trial of the in-control hypothesis, the number of samples required to detect a shift, *M*, is distributed geometric (1-β), where (1-β) is the probability of detecting the shift. If there are M samples, then M-1 failed to detect the shift, and the Mth resulted in detection.
Therefore

$$p(M = m) = (1-\beta)\,\beta^{m-1} \qquad (4.4)$$

which is the distribution for the number of samples taken during the out-of-control period for a Shewhart control chart.

Determining the distribution for the time between the shift and the next sample, *R*, is the next step in the development of the distribution of T. Figure 4.2 shows that R = h-Y, where Y (0 ≤ Y≤ h) is the time between the shift and the preceding sample. Since the time between shifts in the process is distributed exponentially, Y is distributed as a truncated exponential random variable and

$$f_Y(y) = \frac{\lambda e^{-\lambda y}}{1-e^{-\lambda h}},\ 0 \le y \le h.$$

Since R = h-Y,

$$F_R(r) = p(R \le r) = p(h\text{-}Y \le r) = 1 - p(Y \le h\text{-}r).$$

Therefore

$$f_R(r) = \frac{\lambda e^{-\lambda(h-r)}}{1-e^{-\lambda h}}, \quad 0 \le R \le h. \tag{4.5}$$

Convolving (4.4) and (4.5) yields the distribution of T.

$$F_T(t) = p(T \le t) = p(R + M - 1)h \le t)$$

$$= \sum_{m=1}^{\lfloor t/h \rfloor+1} p(M = m)p(R \le t-(m-1)h)$$

$$= \sum_{m=1}^{\lfloor t/h \rfloor} p(M = m) +$$

$$p(R \le t - \lfloor t/h \rfloor h)p(M = \lfloor t/h \rfloor + 1) \tag{4.6}$$

$$= (1-\beta^{\lfloor t/h \rfloor})\frac{1-e^{-\lambda(h-t+\lfloor t/h \rfloor h}}{1-e^{-\lambda h}}(1-\beta)\beta^{\lfloor t/h \rfloor} \tag{4.7}$$

Equation (4.6) results from the definition of cumulative probabilities. If ($T \le R + (M-1)h$) then we sample at least $M = \lfloor t/h \rfloor$ times and detect the shift on the $\lfloor t/h \rfloor + 1^{st}$ sample. Therefore we sum the probabilities that the number of samples ranges from 1 through $\lfloor t/h \rfloor$ and add the probability that

$R \leq t - \lfloor t/h \rfloor h$ times the probability that it takes $\lfloor t/h \rfloor + 1$ samples to detect the shift.

The distribution of Q is derived from the distribution of T through the transformation $Q = (T+nE)\Theta$.

$$F_Q(q) = p((T+nE)\Theta \leq q)$$
$$= p(T \leq \frac{q}{\Theta} - nE).$$

Therefore we can substitute $t = \frac{q}{\Theta} - nE$ into equation (4.7) to obtain the cumulative density of Q.

4.4.2 Derivation of Optimal Buffer Quantity

After a shift occurs we will have produced the items in the buffer and the product at two different defect rates. The distribution of items between the buffer and the product is a function of the quantity produced after the shift, Q, and the buffer size, B. Table 4.1 lists the new variables and parameters required for the buffer problem.

If $Q < B$, production is less than the buffer size, then the buffer contains B - Q units produced at the in-control rate plus Q units produced at the out-of-control rate. If $Q \geq B$, production after the shift is greater than the buffer size, then the buffer contains B units produced at the out-of-control defect rate. Since $Q \geq B$, we will have assembled Q - B units produced at the out-of-control defect rate into the product.

Define

$$\begin{aligned} NB &= \text{number in buffer produced at out-of-control level} \\ &= \min(Q, B) \end{aligned}$$

$$\begin{aligned} E(NB) &= E(\min(Q, B)) \\ &= \int_0^B Qf(Q)dQ + \int_B^\infty Bf(Q)dQ \\ &= E(Q) - \int_B^\infty (Q - B)f(Q)dQ \end{aligned}$$

$$NP = \text{number in product produced at out}-\text{of}-\text{control level}$$
$$= \max(0, Q - B)$$

$$E(NP) = E(\max(0, Q - B))$$
$$= \int_{B}^{\infty} (Q - B) f(Q) dQ$$

Because we can inspect and repair the items in the buffer and in the product, we have twice as many cost and time parameters as in the model of Chapter II. If we let the letters B and P indicate parameters for the buffer and product, and the letters I and R stand for inspection and repair, respectively, then we can define the following notation:

CIB = unit inspection cost in buffer,
CIP = unit inspection cost in product,
ΔI = CIP-CIB,
CRB = unit repair cost in buffer,
CRP = unit repair cost in product,
ΔR = CRP-CRB,
H = cost per cycle of holding B units in inventory

The expected cost per cycle of holding the additional B units in inventory, H, is a function of the firm's cost of capital, i %/hr, the value of a unit \$c/unit, the expected cycle time in hours/cycle and the size of the buffer, B units. We are holding B units at a cost of $\frac{\$ic}{\text{hr unit}}$ for a period of E(cycle time)hours/cycle. Thus

$$H = i\,c\,E(\text{cycle time})\,B$$
$$= H'B \qquad (4.8)$$

Since i, c and E(cycle time) are positive, equation (4.8) is convex in B. The expected savings from holding a buffer arise from the inspection and repair cost differentials, ΔI and ΔR, respectively. The inspection savings is

$$P(\text{process shift is detected})\ \Delta I\ B = (1-\beta)\ \Delta I\ B \tag{4.9}$$

because we inspect the entire buffer only when we detect a shift. For the purposes of computing the expected repair savings, we realize the repair cost differential, ΔR, only on E(NB) units, the expected number of buffer units produced at the high defect rate. The expected repair savings is

$$p(\text{process shift is detected})\ \Lambda_1\ \Delta R\ E(NB)$$

$$= (1-\beta)\ \Lambda_1\ \Delta R\ [\ E(Q) - \int_B^{\infty} (Q-B)\ f(Q)\ dQ] \tag{4.10}$$

Table 4.1 Additional Variables and Parameters for Buffer Study

Variable / Parameter	Symbol
Number of units produced after shift	Q
Size of buffer	B
Unit inspection cost in buffer	CIB
Unit inspection cost in product	CIP
Inspection cost differential	ΔI
Unit repair cost in buffer	CRB
Unit repair cost in product	CRP
Repair cost differential	ΔR
Number in buffer at high rate	NB
Number in product at high rate	NP
99th percentile production quantity	$Q_{.99}$
99th percentile number in product	$NP_{.99}$
Inspection cost	IC
Repair cost	RC
Inventory holding cost / cycle	H
Savings due to buffer	Π
Cost of capital %/hr	i
Value of 1 unit	c

Combining (4.9) and (4.10) yields the savings due to the buffer

$$\Pi = (1\text{-}\beta)\ \Delta I\ B + (1\text{-}\beta)\ \Lambda_1\ \Delta R\ [\ E(Q) - \int_B^{\infty} (Q\text{-}B)\ f(Q)\ dQ]. \tag{4.11}$$

Since $(1\text{-}\beta)$, ΔI, Λ_1 and ΔR are positive and E(NB) is non-decreasing in *B*, equation (4.11) is convex. With the cost and benefits of holding a buffer inventory defined, we next develop the optimality conditions for *B*.

4.4.3 Optimality Conditions for Buffer Size

The optimal buffer size is that B^* for which the marginal savings of increasing the buffer by one item is equal to the marginal cost of increasing the buffer by one item. To determine B^*, we set equal the partial derivatives of H and Π with respect to B and solve for B^*.

$$H = H'\ B$$

$$\frac{\partial h}{\partial B} = H' \tag{4.12}$$

$$\Pi = (1\text{-}\beta)\ \Delta I\ B + (1\text{-}\beta)\ \Lambda_1\ \Delta R\ [E(Q) - \int_B^{\infty} (Q\text{-}B)\ f(Q)\ dQ]$$

$$\frac{\partial \Pi}{\partial B} = (1\text{-}\beta)\ \Delta I + [(1\text{-}\beta)\ \Lambda_1\ \Delta R]\ [1\text{-}F_Q(B)] \tag{4.13}$$

Setting (4.12) equal to (4.13) and solving for *B* yields

$$F_Q(B^*) = 1 - \frac{H' - (1\text{-}\beta)\ \Delta I}{(1\text{-}\beta)\ \Lambda_1\ \Delta R} \tag{4.14}$$

The optimal buffer size is that B^* such that the cumulative probability of Q satisfies equation (4.14).

Several observations can be made from studying equation (4.14) without actually solving it.

1. If we change the 1 in the right side of (4.14) to $\frac{(1\text{-b})\ L1\ DR}{(1\text{-b})\ L1\ DR}$ then the entire right side becomes

$$\frac{(1\text{-b})\ L1\ DR}{(1\text{-b})\ L1\ DR} - \frac{H' - (1\text{-b})\ DI}{(1\text{-b})\ L1\ DR}$$

which is less than or equal to zero if

$$H' \geq (1\text{-b})\ L1\ DR + (1\text{-b})\ DI\ .$$

In this case we hold no buffer because FQ(B*) ≤ 0 if B* = 0. This says that if the cost of holding a unit in the buffer is greater than the expected savings per unit in the buffer, do not hold a buffer.

2. If H' ≤ (1-b) DI, then the right side of (4.14) is greater than 1 and B* = ∞. If the holding cost per unit is less than the expected inspection savings per unit, then the buffer should be infinitely large.

3. As $\Delta R \rightarrow 0, \frac{H' - (1\text{-}\beta)\ \Delta I}{(1\text{-}\beta)\ \Lambda_1\ \Delta R} \rightarrow \infty$ and $1 - \frac{H' - (1\text{-}\beta)\ \Delta I}{(1\text{-}\beta)\ \Lambda_1\ \Delta R}$ becomes negative, thus $B^* \rightarrow 0$. As the repair cost differential decreases, there is less benefit in holding a buffer.

4.4.4 Definition of Out-of-Control Cost

When a shift occurs, we will inspect a predetermined number of items in the buffer and product. Since Q, the number of units produced after the shift, is a random variable, the number of units inspected should be large enough that the probability of Q exceeding it is small. Define $Q_{.99}$ as that number of units such that $p(Q \leq Q_{.99}) = .99$. We will then inspect $Q_{.99}$ units every time a shift occurs. We also define $NP_{.99}$ = max(0,

$Q_{.99}$ -B) to be the number of units inspected in the product after a shift occurs.

The out-of-control cost, C1 , in the cost model of Chapter II can be modified to reflect these buffer-related costs. Let IC be the inspection cost after a shift occurs. Then

$$IC = B * CIB + NP_{.99} * CIP. \tag{4.15}$$

The expected repair cost, E(RC), is

$$E(RC) = \Lambda_1\ E(NB)\ CRB + \Lambda_1\ E(NP)\ CRP \tag{4.16}$$

where Λ_1 is the out-of-control defect rate.

The hourly cost of defects when operating out of control, C1 , is the sum of equations (4.15) and (4.16),

$$C1 = IC + RC.$$

Because we inspect and repair all $Q_{.99}$ items after we have detected a shift, we need only these immediate costs in the model. We have incorporated the E[out-of-control time] equation (2.2) into the new definition for C1, so that term no longer appears in the cost model. The differences between the original cost model, equation (2.9), and the buffer model are the out-of-control cost, C1, and the inventory holding cost H. The new cost model is as follows:

$$C = \frac{\text{in-control+out-of-control+false+real+sampling+holding}}{\text{cycle time}}$$

$$C = \frac{\frac{C_0}{\lambda} + C_1 + \frac{sY}{ARL1} + W}{\frac{1}{\lambda} + \frac{(1-\delta_1)sT_0}{ARL1} - \tau + nE + h(ARL2) + T_1 + T_2}$$

$$+ \frac{\frac{(a+bn)}{h}[\frac{1}{\lambda} - \tau + nE + h(ARL2)] + H}{\frac{1}{\lambda} + \frac{(1-\delta_1)sT_0}{ARL1} - \tau + nE + h(ARL2) + T_1 + T_2} \tag{4.17}$$

4.5 SOLUTION OF THE REVISED COST MODEL

We can extend the algorithm for the c chart of Section 3.3 to the optimization of equation (4.17). We find the optimal buffer size, B^*, by using equation (4.14) after values for n, L and h have been set. We eliminate the approximation of a starting value for h because the values of equation (3.2) depend on C_1, but we cannot determine C_1 until we know a value of B. We modified the c chart algorithm of Section 3.3 to compute the optimal control chart variables and buffer size, and a flowchart of the algorithm is in Figure 4.3.

We applied this algorithm to the same set of parameters used in Chapter III for the low quality system. We set the cost differential so the cost of inspection and repair in the product was two and three times as large as in the buffer. Table 4.2 lists the parameters for this hypothetical production system whose optimal control chart variables appear in Table 4.3. The interpretation of these designs is the same as that in Chapter III with the addition of the buffer being the number of items held in the buffer inventory.

The designs in Table 4.3 have the same sample size and sensitivity level as the c Chart design in Table 3.3. The sampling interval in Table 4.3 is .54 hours shorter than in Table 3.3. The cost of these buffer designs is higher than the design of Table 3.3 because of the higher inspection and repair costs for items in the product. Table 4.3 shows that the chart design and buffer size stay constant as the cost multiplier increases for a given parameter set. But, the cost of the design rises as the cost multiplier increases inspection and repair costs.

4.6 SUMMARY

This chapter shows how we integrate statistical quality control into the design of production systems. We modify the cost model from Chapter II to determine the size of an inventory buffer held between a production and assembly stage. The inspection and repair cost savings offset the holding cost of the additional inventory. We develop the distribution for the number of units produced after a shift in process quality. We use a marginal analysis to determine the optimal buffer quantity. A numerical example demonstrates the effectiveness of the revised model.

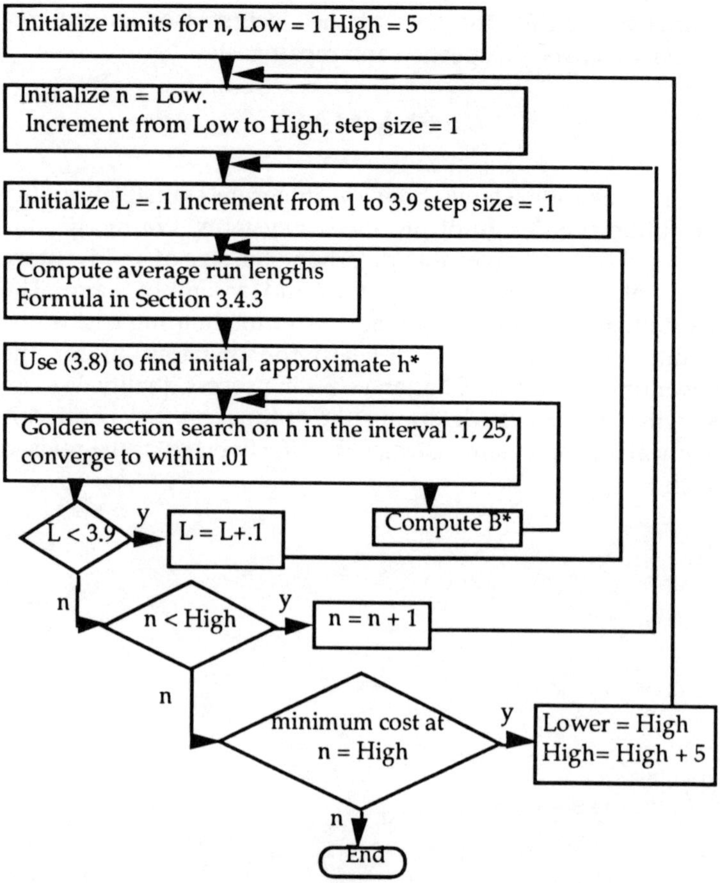

Figure 4.3 c Chart With Buffer Flowchart

Table 4.2 Parameter Levels for Buffer Study Example

PARAMETERS		VALUES
Times (hrs)		
Between shifts	$\frac{1}{\lambda}$	100
Sample/chart 1 item	E	.083
False search time	T_0	.083
Real search time	T_1	.1
Real repair time	T_2	.5
Costs ($)		
Repair of a defect	D	10
False alarm	Y	200
Real alarm	W	50
Fixed sampling	a	1
Variable sampling	b	4
Cost of item	c	100
Miscellaneous		
Magnitude of shift (std dev)	Δ	1
Defects per unit in-control	Λ_0	.1
Production during search	δ_1	0
Production during repairs	δ_2	0
Production rate (units / hr)	Θ	100
Cost of capital (%/hr)	i	.000096
Cost multiplier		2 and 3

Table 4.3 Buffer Solution for c Chart at Low Quality

	Cost Multiplier	
VARIABLE	2	3
n	4	4
L	2.5	2.5
h	2.0	2.0
B	133	133
RESULTS		
ARL1	16.25	16.25
ARL2	2.02	2.02
COST($/hr)	232.22	333.72

V. Conclusions And Recommendations

5.1 OVERVIEW

This chapter summarizes the book and reports the major findings of our research. We discuss the significance of these findings and introduce several areas of future research.

5.2 SUMMARY

This book begins with an introduction of statistical quality control concepts and a motivation for the economic design of quality control charts. We also discuss the relationship between our work and the allocation of inspection effort problem.

Chapter II describes the c, GMA and CUSUM quality control charts and shows how the three charts relate to one another through the weighting factor of the GMA chart. This chapter includes a review of the literature on the economic design of quality control charts and the computation of average run lengths. We also develop a model for the hourly cost of quality control.

In Chapter III we examine the convexity of the cost model and conclude that the model is convex in sampling interval and sensitivity factor. The model appears to be convex in sample size, but we cannot prove it. We describe our algorithms for the minimization of the cost model and show that we are able to obtain a lower cost solution than that published in a prior work. We describe the calculation of average run lengths for the c, GMA and CUSUM charts. We also present numerical examples, comparisons of cost performance for the three charts, and a sensitivity analysis. The chapter concludes with an analysis of the effects of process shutdown on the model's solution.

In Chapter IV we extend the economic model developed in Chapter II to the design of production systems. We discuss the relationship of our buffer system to manufacturing systems which use inventory buffers to decouple operations. The revised model computes the control chart variables and determines the size of an inventory buffer between a production and an assembly operation. This buffer reduces the cost of quality when it is less expensive to inspect and repair items before assembly. We develop the distribution of the quantity produced after a shift in quality occurs and derive optimality conditions for the buffer inventory. These enable us to show under which conditions there should be no buffer and when the buffer should be infinitely large. We add this buffer size calculation to the c chart algorithm of Chapter III, and a numerical example shows the use of the revised cost model.

5.3 CONCLUSIONS

We have shown that the economic modeling of statistical quality control charts can be extended to low defect systems. The data from our numerical studies show that the Shewhart c chart is less expensive at low quality levels where the shift in quality is relatively large. As the quality of the process improves, the CUSUM chart becomes significantly less expensive to use than the c or GMA charts. These higher quality systems encounter smaller shifts in quality than the low quality systems. In Section 2.2.4, we explain that the c chart is more effective than the CUSUM or GMA charts at detecting large shifts in quality, while the converse is true for small shifts in quality. Our findings support these observations about the effectiveness of the three control charts.

While our numerical findings support the use of c charts for large shifts and CUSUM charts for small shifts, these results are for one set of data. To be certain of finding the minimum cost control chart for a specific process, we should apply the three algorithms to the parameters of that process.

In Section 3.6, we show that there are significant savings possible from improved process quality. We recommend that users rely on the cost model in process selection decisions. Managers can compute the cost of quality for a process and factor this information into their financial analysis. These savings in quality costs can help justify the purchase of higher reliability manufacturing systems.

The economic model contains two parameters which show that processing stops during the search for an assignable cause and the repair of the process. By changing the values (0,1) of these parameters, we can compare four different production control policies. We can allow the process to continue to operate after a shift is detected, which increases the number of non-conforming units produced but reduces the number of process set-ups. Or we can stop the process every time an out-of-control condition is indicated which increases set-up costs but reduces the number of non-conforming units to inspect and repair. This is a valuable tool for managers who have the flexibility of implementing several production control policies.

In our numerical examples, we found that there were differences of as much as 11 percent between the costs of different policies for low-quality systems. However, the cost differences are insignificant in high-quality systems. This occurs because a high-quality system produces only occasional defects, and shifts in process quality occur very infrequently. Therefore, there will be very few defects produced if the process continues to operate when it experiences one of its rare shifts in quality. Users will seldom incur additional set-up costs and their cost will be spread over the long cycle time of a high quality system.

We show that the optimal control chart variables are robust to errors in cost parameter estimation. The economic designs are as much as 45 percent less expensive than a heuristic design. Managers have been reluctant to use the economic design of control charts because they were uncertain of their cost estimates. Our findings are significant because they show that the economic designs are relatively insensitive to errors in parameter values, and they yield large savings over the traditional heuristic designs. This should encourage managers to accept the idea of the economic design of control charts.

The most significant contribution of this research is the integration of the economic design of control charts with the design of production systems. This is the first work which addresses the disposition of post-production items based on the quality of the manufacturing system and the costs of inspecting and repairing the item. We show that managers can reduce the cost of quality for a process by including a buffer between the production and assembly stages. This tends to contradict the modern production philosophies of JIT and Kanban which recommend the use of very small buffer inventories. We develop the distribution function of the number of units produced after a shift occurs. Using marginal analysis, we find the optimality conditions for the buffer inventory.

5.4 EXTENSIONS

This research on the design of quality control charts can be extended in several directions. The economic model assumes that the time until the shift occurs is distributed exponentially. This is appropriate where the probability of a shift is independent of the length of time the process has operated. A process which exhibits wear over time is more appropriately modeled by a distribution which reflects this increasing probability of failure. This is an important extension because many machining operations, for example, those which experience tool wear, would finally be able to use control charts based on a minimum cost design.

In many manufacturing systems, the technology exists to inspect every unit produced. Manufacturers accomplish this with automatic scales, laser scanners, and high speed electronic test equipment, among others. This equipment allows us to detect all defective or non-conforming items and record the time between their occurrences. Process quality can be monitored by observing the time between defects. We assume that the process is operating in control as long as the time between defects is greater than τ, where τ is the decision variable. If the inter-defect time is less than or equal to τ, the process is stopped and repaired, if necessary. We want to develop a model which determines the optimal value for τ based on the costs of a defective item, process shutdown, and process repair.

Some production processes require the control of more than one quality characteristic. Studies show some work on control charts for multiple measurement data but no one has published findings on the control of multiple types of defects. We need to define this new multi-defect control system and develop an economic model to determine the optimal control variables.

The sections on the allocation of inspection effort and buffer inventories show that we can integrate statistical quality control into these problems to build a more complete model of the production process. Such a model would consider the quality of incoming components, the output quality of each stage, and any shifts in process quality. The objective would be to design all of the control charts and determine the buffer inventories for an entire production line. This is particularly relevant given the interest in flexible manufacturing systems which contain several interdependent machines and require the determination of buffer inventory capacities.

In conclusion, manufacturing technology is producing more reliable production systems which require new methods of production and quality control. We have investigated two relevant topics and identified several new ones related to this changing technology.

Bibliography

Aroian, L. and Levine, H. (1950), "The Effectiveness of Quality Control Charts," *Journal of the American Statistical Association*, 44, 252

Barish, N.N. and Hauser, N. (1971), "Economic Design for Control Decisions," *Journal of Industrial Engineering*, 14, 3

Britney, R.R. (1972), "Optimal Screening Plans for Nonserial Production Systems," *Management Science*, 18, 9

Brook, D. and Evans, D.A. (1972), "An Approach to the Probability Distribution of CUSUM Run Length," *Biometrika*, 59, 3

Buzacott, J. and Hanifin, L.E. (1978), "Models of Automatic Transfer Lines with Inventory Banks: A Review and Comparison," *AIIE Transactions*, 10, 197

Chiu, W.K. (1974), "The Economic Design of CUSUM Charts for Controlling Normal Means," *Applied Statistics*, 23, 3

Chiu, W.K. (1975), "Economic Design of Attribute Control Charts," *Technometrics*, 17, 1

Chiu, W.K. and Wetherhill, G.B. (1974), "A Simplified Scheme for the Economic Design of $\bar{x}$-Charts," *Journal of Quality Technology*, 6, 2

Cowden, D.J. (1957), *Statistical Methods in Quality Control*, Prentice-Hall, Englewood Cliffs, NJ

Crosby, P.B. (1979), *Quality is Free*, Mentor, New York, NY

Crosier, R.B. (1963), "A New Two-Sided Cumulative Sum Quality Control Scheme," *Technometrics*, 28, 3

Crowder, S.V. (1986), "Average Run Lengths of Exponentially Weighted Moving Average Control Charts," *Journal of Quality Technology*, 19, 3

Davis, D.J. (1952), "An Analysis of Some Failure Data," *Journal of the American Statistical Association*, 47, 258

Deming, W.E. (1986), *Out of the Crisis*, MIT/CAES, Cambridge, MA

Duncan, A.J. (1956), " The Economic Design of $\bar{x}$ Charts Used To Maintain Current Control of a Process," *Journal of the American Statistical Association*, 51, 274

Duncan, A.J. (1974), *Quality Control and Industrial Statistics*, Irwin, Homewood, IL

Epstein, B. (1958), "The Exponential Distribution and Its Role in Life-Testing," *Industrial Quality Control*, 15, 6

Gardiner, J.S. (1987), *Detecting Small Shifts in Quality Levels in a Near Zero Defect Environment for Integrated Circuits*, Unpublished Ph.D. Dissertation, University of Washington, Seattle, WA

Girshick, M.A. and Rubin, H. (1952), "A Baye's Approach to a Quality Control Model," *Annals of Mathematical Statistics*, 23, 1, 114

Goel, A.L. and Wu, S.M. (1971), "Determination of ARL and a Contour Nomogram for CUSUM Charts to Control Normal Means," *Technometrics*, 13, 2

Goel, A.L. and Wu, S.M. (1973), "Economically Optimum Design of CUSUM Charts," *Management Science*, 19, 11

Goldsmith, P.L. and Whitfield, H. (1961), "Average Run Lengths in Cumulative Chart Quality Control Schemes," *Technometrics*, 3, 1

Grant, E.L. and Leavenworth, R.S.(1980), *Statistical Quality Control*, McGraw Hill, New York, NY

Hunter, J.S.(1986), "The Exponentially Weighted Moving Average," *Journal of Quality Technology*, 18, 4

Johnson, N.L. and Leone, F.C. (1962), "Cumulative Sum Control Charts - Mathematical Principles Applied to Their Construction and Use," *Industrial Quality Control*, 18, June, July, August

Juran, J.M. and Gryna, F.M. (1980), *Quality Planning and Analysis*, McGraw Hill, New York, NY

Kemp, K.W. (1961), "The Average Run Length of the Cumulative Sum Chart When a V-mask Is Used," *Journal of the Royal Statistical Society* (B), 23, 149

Khan, R.A. (1984), "On Cumulative Sum Procedures and the SPRT with Applications," *Journal of the Royal Statistical Society (B)*, 46, 1

Knappenberger, H.A. and Grandage, A.H.E. (1969), "Minimum Cost Quality Control Tests," *AIIE Transactions*, 1, 1

Lindsay, G.F. and Bishop, A.B. (1964), "Allocation of Screening Inspection Effort - A Dynamic Programming Approach," *Management Science*, 10, 2

Lorenzen, T.J. and Vance, L.C. (1986), "The Economic Design of Control Charts: A Unified Approach," *Technometrics*, 28, 1

Lucas, J.M. (1976), "The Design and Use of V-Mask Control Schemes," *Journal of Quality Technology*, 8, 1

Lucas, J.M. (1985), "Counted Data CUSUM's," *Technometrics*, 27, 2

Montgomery, D.C. (1980), "The Economic Design of Control Charts: A Review and Literature Survey," *Journal of Quality Technology*, 12, 2

Montgomery, D.C. (1984), *Design and Analysis of Experiments*, Wiley, New York, NY

Montgomery, D.C. (1985), *Introduction to Statistical Quality Control*, Wiley , New York, NY

Okamura, K. and Yamashina, H. (1979), "Justification for Installing Buffer Stocks in Unbalanced Two Stage Automatic Transfer Lines," *AIIE Transactions*, *11*, 4, 308

Page, E.S. (1954), "Continuous Inspection Schemes," *Biometrika*, 41, 100

Page, E.S. (1961), "Cumulative Sum Charts," *Technometrics*, 3, 1

Page, E.S. (1963), "Controlling the Standard Deviation by CUSUM's and Warning Lines," *Technometrics*, 5, 3

Panagos, M.R., Heikes, R.G. and Montgomery, D.C. (1985), "Economic Design of $\bar{x}$ Control Charts for Two Manufacturing Process Models," *Naval Research Logistics Quarterly*, 32, 631

Roberts, S.W. (1959), "Control Chart Tests Based on Geometric Moving Averages," *Technometrics*, 1, 3

Roberts, S.W. (1966), "A Comparison of Some Control Chart Procedures," *Technometrics*, 8, 411

Robinson, P.B. and Ho, T.Y. (1978), "Average Run Lengths of Geometric Average Charts by Numerical Methods," *Technometrics*, 20, 1

Ross, S.M. (1970), *Applied Probability Models with Optimization Applications*, Holden-Day, San Francisco, CA

Saniga, E.M. and Shirland, L.E. (1977), "Quality Control in Practice - A Survey," *Quality Progress*, 10

Schonberger, R.J. (1982), *Japanese Manufacturing Techniques*, The Free Press, New York, NY

Shewhart, W.A. (1926 a), "Finding Causes of Quality Variations," *Manufacturing Industries*, February, 125

Shewhart, W.A. (1926 b), "Quality Control Charts," *Bell Systems Technical Journal*, October, 593

Taylor, H.M. (1968), "The Economic Design of Cumulative Sum Control Charts," *Technometrics*, 10, 3

Vance, L.C. (1983), "A Bibliography of Statistical Quality Control Chart Techniques, 1970-1980," *Journal of Quality Technology*, 15, 2

Vardeman, S. and Ray, D. (1985), "Average Run Lengths for CUSUM Schemes When Observations are Exponentially Distributed," *Technometrics*, 27, 2

Waldmann, K.H. (1986), "Bounds for the Distribution of the Run Length of One-Sided and Two-Sided CUSUM Quality Control Schemes," *Technometrics*, 28, 1

Weiler, H. (1952), "On the Most Economical Sample Size for Controlling the Mean of a Population," *Annals of Mathematical Statistics*, 23

White, L.S. (1968), "The Analysis of a Simple Class of Multistage Inspection Plans," *Management Science*, 12, 9

Woodall, W.H. (1983), "The Distribution of the Run Length of One-Sided CUSUM Procedures for Continuous Random Variables," *Technometrics*, 25, 3

Woodall, W.H. (1984), "On the Markov Chain Approach to the Two-Sided CUSUM Procedure," *Technometrics*, 26, 1

Woodall, W.H. (1986), "The Design of CUSUM Quality Control Charts," *Journal of Quality Technology*, 18, 2

Yum, B.J. and McDowell, E.D. (1981), "The Optimal Allocation of Inspection Effort in a Class of Nonserial Production Systems," *AIIE Transactions*, 13, 4

Index